Ryan King

Giant Pandas as a Model Species for Extinct Hominin Genus Paranthropus

Ryan King

Giant Pandas as a Model Species for Extinct Hominin Genus Paranthropus

An Analysis of the Mechanical Properties of Bamboo and Their Relationship to the Diet of the Hominin Genus Paranthropus

LAP LAMBERT Academic Publishing

Impressum / Imprint

Bibliografische Information der Deutschen Nationalbibliothek: Die Deutsche Nationalbibliothek verzeichnet diese Publikation in der Deutschen Nationalbibliografie; detaillierte bibliografische Daten sind im Internet über http://dnb.d-nb.de abrufbar.
Alle in diesem Buch genannten Marken und Produktnamen unterliegen warenzeichen-, marken- oder patentrechtlichem Schutz bzw. sind Warenzeichen oder eingetragene Warenzeichen der jeweiligen Inhaber. Die Wiedergabe von Marken, Produktnamen, Gebrauchsnamen, Handelsnamen, Warenbezeichnungen u.s.w. in diesem Werk berechtigt auch ohne besondere Kennzeichnung nicht zu der Annahme, dass solche Namen im Sinne der Warenzeichen- und Markenschutzgesetzgebung als frei zu betrachten wären und daher von jedermann benutzt werden dürften.

Bibliographic information published by the Deutsche Nationalbibliothek: The Deutsche Nationalbibliothek lists this publication in the Deutsche Nationalbibliografie; detailed bibliographic data are available in the Internet at http://dnb.d-nb.de.
Any brand names and product names mentioned in this book are subject to trademark, brand or patent protection and are trademarks or registered trademarks of their respective holders. The use of brand names, product names, common names, trade names, product descriptions etc. even without a particular marking in this works is in no way to be construed to mean that such names may be regarded as unrestricted in respect of trademark and brand protection legislation and could thus be used by anyone.

Coverbild / Cover image: www.ingimage.com

Verlag / Publisher:
LAP LAMBERT Academic Publishing
ist ein Imprint der / is a trademark of
OmniScriptum GmbH & Co. KG
Heinrich-Böcking-Str. 6-8, 66121 Saarbrücken, Deutschland / Germany
Email: info@lap-publishing.com

Herstellung: siehe letzte Seite /
Printed at: see last page
ISBN: 978-3-659-61475-0

ACKNOWLEDGEMENTS

The author wishes to acknowledge the work of Gianna Covelli for assisting with data collection. The help of Mark Wagner was essential in fixing software bugs and providing insight into a method of data collection. Thank you to the staff of the Department of Biological Science for their assistance and support.

CONTENTS

LIST OF TABLES

LIST OF FIGURES

ABSTRACT

Giant pandas (*Ailuropoda melanoleuca*) have cranial morphology similar to the extinct hominin genus *Paranthropus* which makes them an excellent model species when studying *Paranthropus* diet. Both species have wide skulls with flared zygomatic arches adapted for chewing. To gain insight into possible food sources of *Paranthropus*, I investigated the giant panda's specialized diet of bamboo. The toughness, hardness, and stiffness of various bamboo species was determined to assess mechanical challenges facing giant pandas during feeding. Bamboo is thought to be tough, but studies on such properties and how they apply to mastication of giant pandas are largely absent from the scientific literature. Knowing the properties of bamboo will help draw a parallel between giant panda and *Paranthropus* diets. Mechanical properties data were gathered from young and adult bamboo shoots using a universal testing machine, which applies and measures force to the bamboo samples. A collection of four species, which include bamboo favored and ignored by giant pandas, were tested to determine how bamboo properties vary interspecifically with the goal of discovering if there are mechanical differences between bamboo favored and disliked by the species. Conducting this research will aid efforts to understand the diet of *Paranthropus* and could help establish a link between *Paranthropus* and a food source with properties comparable to those of bamboo.

CHAPTER 1

INTRODUCTION

The giant panda (*Ailuropoda melanoleuca*) has skull morphology similar to that of the extinct hominin genus *Paranthropus,* which could make it a useful model species for reconstructing the diet of these early hominins. To gain insight into possible food sources of *Paranthropus*, this study will look at the giant panda's diet of bamboo. In order to understand the kinds of stress involved in the mastication of bamboo, details about bamboo's mechanical properties must be made available. Little is known about these properties in bamboo and less is known about the properties of bamboos used by giant pandas as food material. This study seeks to fill in the gaps in the literature by providing data on the toughness, hardness, and Young's modulus (stiffness) of several species of bamboo in order to assess mechanical challenges facing the giant panda during feeding.

By some estimates, bamboo makes up approximately 99% of the giant panda's diet (Wei *et al.*, 1999). Therefore, the difficult mastication of bamboo is thought to be the driving force behind the giant panda's derived masticatory morphology. This leads to the expectation that bamboo will have relatively high toughness, hardness, and/or stiffness. If this is assumed to be the case, then perhaps a similar selective pressure was the force behind the masticatory development of the genus *Paranthropus*. *Paranthropus* is an extinct genus of hominin (humans and our ancestors) whose cranial morphology closely mirrors that of the giant panda (Davis, 1964; Du Brul, 1977). The skulls of both giant panda and *Paranthropus* share key characteristic features which are linked to mastication. Much like how the giant panda is morphologically derived among bears (Sacco & van Valkenburgh, 2004), *Paranthropus* too is derived among early hominins (Constantino and Wood, 2007). If it can be accepted that convergent masticatory morphology is a reflection of similar mechanical demands being placed on the skull, then learning more about dietary habits of giant pandas could help establish a link between *Paranthropus* and a food source with properties comparable to those of bamboo.

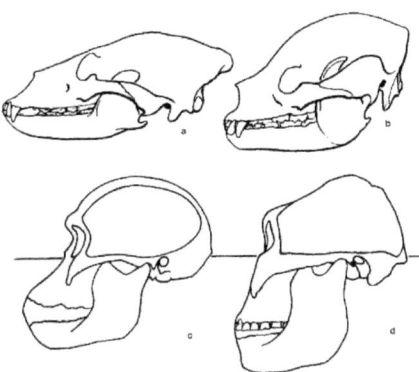

Figure 1. Comparison of Giant Panda and Paranthropus *to Similar Species. Taken from Du Brul, 1977. This figure shows the unique cranial specialization of both* Ailuropoda melanoleuca *and* Paranthropus *compared to a closely related member of their respective group. From the top left, skull "a" shows the morphology of a brown bear and on the right, the skull of a giant panda is marked "b". Note the skull of the panda is more orthognathic (retraction of the face) with deeper jaws and larger molars. The bottom left is a skull of* Australopithecus africanus *(marked "c") compared to* Paranthropus boisei *(marked "d") on the right. Note again the similar features to the giant panda.*

PROBLEM STATMENT

The hardness and toughness of bamboo is thought to be the driving force behind the specialized cranial and masticatory adaptions of *Ailuropoda melanoleuca* (Christiansen, 2007). However, little information is available in the scientific literature regarding bamboo's mechanical properties and how they relate to giant panda mastication. Research has found certain grasses to be tough (Kobayashi *et al.*, 2008) and because bamboo belongs to a family of grasses, it is likely bamboo is also tough. The skull structure of the giant panda, which seems to be adapted to frequent chewing, generating and/or dissipating high force, is consistent with bamboo being a tough food source. Data will be provided on the mechanical properties of bamboo and fill in some of the knowledge gaps which surround these properties. The properties examined include toughness, hardness, and elastic modulus (stiffness). Information on these will be obtained through the use of a portable universal testing machine (Lucasscientific.com). Although other studies have been performed on bamboo's mechanical properties (Low and Che, 2006), these have focused on the application of bamboo for construction or technological purposes. Our approach attempts to link the mechanical properties of bamboo to the mastication of giant pandas and focuses on species of bamboo native to giant panda habitats.

IMPORTANCE OF DIET AND ITS INFLUENCE ON MORPHOLOGY

Diet is so well engrained in the life of an organism that a change in diet can signify a milestone in the evolution of that organism's lineage (Ungar and Sponheimer, 2011). Diet plays a crucial role in an organism's life cycle and often dictates behavior patterns. Understanding the diet of an organism can shed light on how that organism may have lived. In order to understand the diet of our early hominin ancestors, several methods have been used including comparative and functional morphology. A combination of these techniques makes it possible to gain insight into what kinds of foods our ancestors may have eaten and how they were consumed.

FUNCTIONAL MORPHOLOGY

A potential indicator of foods hominins may have been eating is their functional morphology. The size and structure of the teeth are especially good indicators, particularly the thickness of tooth enamel. Both *Paranthropus boisei* (Grine and Martin, 1988) and to a lesser extent *Paranthropus robustus* (Olejniczac *et al.*, 2008) had thick tooth enamel which may have helped prolong the life of the tooth as it was slowly worn down by day to day use. Another benefit of thick tooth enamel is the potential to resist tooth fracture when biting hard objects. Hard foods can create small areas of high stress when contacting the enamel. Thicker enamel should allow the teeth to withstand greater amounts of stress caused by the mastication of mechanically challenging foods (Lucas *et al.*, 2008). Common hard foods are nuts or seeds which are protected by a

3

fracture resistant shell or covering. When biting these foods, the highest amount of force is generated by the initial bite. Once the teeth cause a fracture, it takes less energy to continue growing the fracture and bite through the food. Interestingly, both hard foods and teeth evolved similar structures for protection (Lucas *et al.*, 2008). A fracture resistant coating is beneficial to both seeds and teeth so both are selected for in Nature. Enamel can serve other purposes beyond wear and fracture resistance. The distribution of the enamel is also important as it can influence which parts of the teeth are worn down first. Some animals use this wear pattern to hone their teeth to a sharp point, sculpting a new tooth shape by wearing down the excess enamel. This practice can be seen in the way goats chew their food, slowly wearing down the extra tooth enamel to form sharp crests useful for slicing through tough vegetation (Lucas *et al.*, 2008).

Foods typically possess one of two varieties of mechanical defense which are stress limited or displacement limited. Stress limited defenses usually involve being strong and stiff, requiring a great amount of force per area to initiate a crack. The drawback to this kind of defense is a tradeoff between hardness and brittleness. It may take a large force to cause the initial fracture, but once that fracture has been made it requires much less force to advance. Organisms using displacement limited defenses are tough and flexible. Little force is required to cause an initial crack, but it is difficult to propagate the crack once it has been started. Certain foods, especially some fruits, have properties that use a mixture of stress and displacement limited defenses (Lucas, 2004; Ungar and Lucas, 2010).

Different tooth shapes can be more effective biting through stress or displacement limited defenses (Lucas, 2004). Animals who exploit stress limited defenses (hard object feeders) typically have blunt and domed molar cusps to concentrate the force of a bite onto a small area to assist with the initial fracture. Organisms that eat foods protected by displacement limited defenses are aided by shear-like crests or blades that can slice through the tough material. These observations are supported by studies of many extant primates that exploit hard or tough materials as fallback foods (Kay and Covert, 1984; Strait, 1993).

Both *Paranthropus* (Du Brul, 1977) and the giant panda (Davis, 1964) have "molarized" premolars which are enlarged to the point of resembling molars. While biomechanics models (Du Brul, 1977; Spencer, 1998) report that premolars are unlikely to be involved in the generation of maximum bite forces, Wood and Strait (2004) have suggested that the enlargement of the premolars may have allowed *Paranthropus* to process a larger volume of food at one time. This suggestion is supported by the findings of Walker (1981), that report that larger tooth size may allow for the faster and more efficient consumption of a given food. Given this information, the enlarged premolars of the giant panda may assist when processing high volumes of bamboo, but are unlikely to be able to generate as much force as the molars.

Along with enlarged, bunodont molars, flared zygomatic arches are also characteristic of both giant pandas and *Paranthropus* (Davis, 1964). The zygomatic bones are wider and more anteriorly positioned in *Paranthropus* than in other hominins (Constantino and Wood, 2007) which could have allowed for the attachment of larger masseter muscles, greater mechanical advantage of those muscles, and a larger passageway for the temporalis muscle in the infratemporal fossa. Larger muscles leveraged for greater mechanical advantage would have allowed for higher bite forces than in other hominins (Demes and Creel, 1988). *Paranthropus* also exhibits a substantial degree of facial orthognathy, or shortening of the face, which is similar to the shortened jaws of the giant panda relative to extant ursines (Fig. 1; Constantino and Wood, 2007; Christiansen, 2007).

Much like *Paranthropus* is differentiated from other hominins by its robust jaws and dentition, the masticatory system of the giant panda makes it unique among ursines (Davis, 1964). As revealed by Christiansen (2007), giant pandas can generate the highest bite forces of all extant bear species. The giant panda owes these high bite forces to the increased areas of attachment of the masseter and temporalis muscles (although it should be noted that giant pandas do not have the highest masseter/temporalis muscle to skull size ratio in ursines), enlarged molars, and wide, flaring zygomatic arches (Christiansen, 2007). While the giant panda may have masticatory adaptations which aid in its consumption of bamboo, its digestive system is not suited to this specialized diet and cannot digest the cellulose and lignin present in herbaceous material (Davis, 1964). Christiansen (2007) remarked that the giant panda is uniquely specialized among ursines and possesses features which appear to be adaptations that arose from the selective pressure of the mastication of bamboo. These adaptations include widened zygomatic arches, a domed skull, and enlarged molars, all of which assist in allowing the giant panda to generate high bite forces relative to its body size. Christiansen and Wroe (2007) conclude that high bite forces relative to body size, along with heavily molarized dentition, are characteristics of an evolutionary trend toward the specialization of mechanically resistant plant material in carnivoran lineages. Adaptations that favor an herbivorous diet, particularly one high in mechanically resistant plants, are markedly distinct from those of other ursines which have more carnivorous or omnivorous diets (Christiansen, 2007). Giant pandas can process bamboo stalks with a diameter of up to an inch and a half (Du Brul, 1977). Because the digestive system of giant pandas cannot fully utilize the nutrients found in bamboo, pandas must continuously consume around 15-20 kg of bamboo per day. This constant mastication is thought to put a large amount of strain on the jaws and teeth of pandas which are presumed to be adapted for handling the stress (Constantino *et al.*, 2007).

Both Davis (1967) and Du Brul (1977) comment that the specialized cranial morphology of *Ailuropoda melanoleuca* bears a resemblance to the cranial features of the extinct hominin genus *Paranthropus*. Both possess flared zygomatic arches and large molars which appear to be adapted to generate high bite forces. As seen in the giant panda, the skull of *Paranthropus* is also specialized (with regards to cranial morphology) among those in its group (Du Brul, 1977). If giant pandas are able to utilize the adaptations which set them apart from other members of their group to consume bamboo, then perhaps *Paranthropus* had used its derived morphology to consume foods that are mechanically similar to bamboo.

A limitation of using functional morphology to infer diet is that specialized adaptations may not accurately reflect the kinds of food preferred by the animal. Somewhat paradoxically, specializations in functional morphology do not necessarily indicate what foods an organism usually ate. Specializations can reflect foods eaten in only the most extreme circumstances and in fact, the species may avoid eating the type of food for which it is has specializations (Ungar, Grine, and Teaford, 2008). This conundrum is often referred to as Liem's Paradox, which refers to a situation where some organisms with morphology indicative of dietary specialization can, in actuality, subsist on a more generalized diet. This paradox is classically associated with cichlid fish species that possess derived feeding mechanisms adapted for particular food items. These fish can often forgo the food source they are adapted for in favor of more common prey (Liem, 1990). While this observation may have initially led to the conclusion that these specialized adaptations do not offer a competitive advantage and only exist in the population because they are not detrimental to the organism's survival, Robinson and Wilson (1998) have presented a model which suggests that morphological adaptations allow an organism to exploit resources which are normally difficult to utilize and do not interfere with the consumption of more generalized food sources. Gathering resources in this way would allow an organism to take advantage of a broader spectrum of foods and therefore optimize time spent foraging.

Morphological specializations should only reflect the most challenging food items the animal eats. Eating softer foods would not require special adaptations for chewing, so even if these foods were selected or even preferred it would not be evident from the morphology. Teeth are often able to resist forces required to breach most foods and their strength is only relevant when trying to infer what foods could have been eaten, not how often challenging foods were consumed (Constantino *et al.*, 2009; Wood and Strait, 2004). Eating challenging foods may only occur when the animal's preferred food source is unavailable, in which case they may switch to a less preferred fallback food. Ordinarily, gorillas and chimpanzees have a high amount of overlap in their diets with each of them preferring to eat soft fruits. Gorillas and chimpanzees living sympatrically have been documented to have a 73% dietary overlap (Ungar, 2004). However during times when preferred fruits are in short supply, chimpanzees will switch to harder foods like nuts or seeds while gorillas will fall back on tough vegetation (Ambrose, 2006). The shearing crests on the teeth of gorillas allow them to more easily process the tough plant material, but are not necessary when eating the fruit on which they usually feed. Chimpanzee molars lack the shearing crests of gorillas and have teeth better suited to crushing or grinding. Note that the tooth morphology is most beneficial when masticating the less favored fallback foods and are less critical when consuming the preferred food source.

BAMBOO AND THE GIANT PANDA

Bamboo is a fast growing evergreen plant in the grass family Poaceae which grows in clumps through the utilization of a rhizome system (McClure, 1993). Bamboo is a composite material consisting of a fibrous outer surface and a largely hollow interior (Low and Che, 2006). The stalk (or culm) makes up the bulk of the plant and is segmented by nodes. The inter culm of woody bamboos are lined with lignified pith which becomes more spongy near the growing tips (Yamashita et al, 2009). New branch shoots arise from the nodes and leaf compliments are formed at the terminal ends of the shoots. Individual bamboo fibers are composed of cellulose and form vascular bundles which can alter the hardness of the culm depending on the arrangement and number of the bundles (Jain *et al*, 1992).

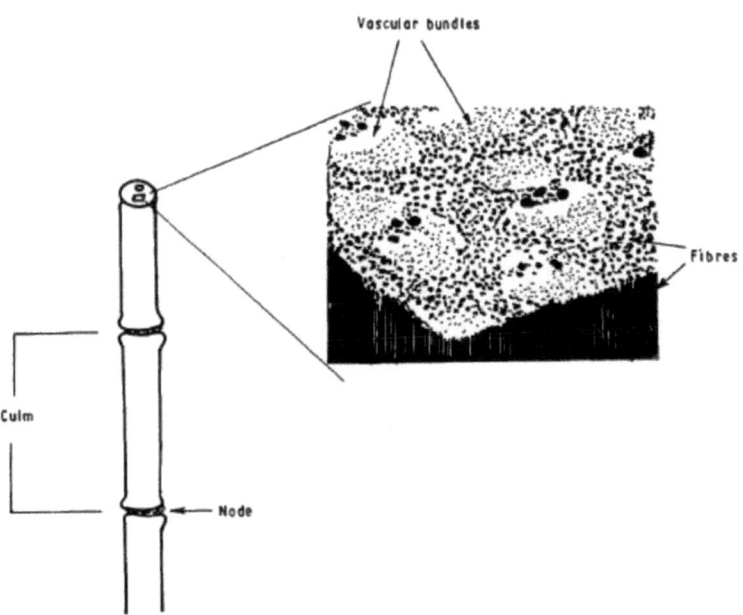

Figure 2. Structure of Bamboo. Taken from Jain et al, 1992. The above figure shows the culm and nodes of bamboo as well as the fiber structure and vascular bundles.

While giant pandas feed on bamboo year round, their utilization of the plant varies depending on the season. Wei *et al*. (1999) documented that giant pandas in Yele Natural Reserve in the Sichuan province of China mainly fed on bamboo stems throughout the months of March and April and shifted their focus to bamboo shoots in

May. Stems of bamboo consist of the culm (Fig. 2) of the bamboo stalks and shoots are young bamboo which eventually form new culms. From July to October, giant pandas would eat the leaves of the bamboo almost exclusively with 92% of their diet consisting of leaves. When feeding on leaves, giant pandas were observed biting off the stems and holding them rather than bending the stems over to get to the leaves. For the remainder of the year until the following March, the panda would forage the stems of old bamboo shoots. The study by Wei *et al.* (1999) also found that giant pandas prefer to eat bamboo shoots which are taller and more robust with a larger diameter. Giant pandas will ignore slimmer bamboo shoots in favor of taller and larger plants. Another study suggests giant pandas prefer to forage on the edges of bamboo patches because the edges contain thicker shoots of bamboo (Yu *et al.*, 2003).

The giant panda, while possessing a number of specialized adaptations for ingesting bamboo, is inefficient at digesting bamboo (Dierenfeld *et al.*, 1982). Studies documenting the digestibility of bamboo by giant pandas discovered the percentage of bamboo able to be digested to be less than 20% (Dierenfeld *et al.*, 1982). The passage of bamboo through the digestive tract is also very rapid. Dierenfeld *et al.* (1982) suggests that while the giant panda is inefficient when digesting bamboo, its specialized masticatory systems may be able to finely chew up the bamboo to increase the amount of nutrients usable by the giant panda (Dierenfeld *et al.*, 1982). A study on the fecal flora of the giant panda revealed a change in fecal bacteria as a young giant panda matured and started feeding on bamboo leaves. A change in fecal flora is also seen as the seasons affect which parts of bamboo giant pandas feed on (Hirayama *et al.*, 1989; Williams *et al.*, 2012).

MATERIAL PROPERTIES OF BAMBOO

Three mechanical properties are determined for bamboo in this study. These properties are toughness, hardness, and Young's modulus. Toughness is a form of mechanical defense which focuses on resisting the spread of cracks rather than preventing cracks themselves (Lucas, 2004). Toughness is measured as the amount of work that is done for a crack to increase in area. Toughness is related to this study because it represents the amount of work an animal must do to masticate its food source

(Turner *et al.*, 1993) and may be directly relevant to how an animal chooses what foods to feed upon (Choong *et al.*, 1992). For this study, toughness is acquired through scissors cut tests using the equation

$$R = \frac{Wc - Wf}{Lt}$$

where R is the toughness of the material, W_c is the work of creating a cut, W_f is the work of friction created by the metal scissor blades passing one another, L is the length of the cut, and t is the thickness of the specimen (Darvell *et al.*, 1996). Because of the variability present in the shape of scissor blades, the length of the cut is measured directly. The work of friction is subtracted from the total work of the cut to accurately report only the resistance of the material.

Hardness, when used scientifically, refers to a resistance to plastic deformation when under stress (Lucas, 2004). Hardness is highly correlated to the yield strength of a material and in cases were the material collapses inward on itself (such as plant material where the cell walls burst, flatten, then compress) the hardness of the material is equal to its yield strength (Lucas, 2004). The most common method of determining hardness is Vicker's indentation where a sharp indenter tip is used to deform the specimen. Using this method, hardness can be mathematically defined as H = F/A where H is hardness, F is force, and A is the area of indentation (Lucas, 2004).

Young's (elastic) modulus is related to the stiffness or rigidity of a material. It is measured as the force producing unit of deformation of a specimen relative to the dimensions of the specimen. Put simply, Young's modulus is a ratio of stress to strain which is measured in units of force per unit area (Lucas, 2004). Force is converted to stress by dividing by the area of the specimen the force acts upon. Strain is found by dividing the original dimensions of the specimen in the direction of the force. This can be mathematically displayed as:

$$E = \frac{Fl}{Al_o}$$

where E is Young's (elastic) modulus, F is the force acting on the specimen, l is instantaneous length of the specimen, A is the area on which the force acts upon, and l_o is the original length of the specimen (Lucas, 2004).

Typical units of modulus are units of pressure (Megapascals (10^6) and Gigapascals (10^9)). Bending tests are commonly used to determine Young's modulus. These tests apply force to a material and measure the displacement caused by the stress (Lucas, 2004).

Although few studies exist on the mechanical properties of bamboo, a study by Low and Che (2006) features results on the toughness, hardness and elastic modulus of bamboo. They found younger bamboo has greater fracture toughness and a higher Young's modulus than older bamboo. They have also found variability in the hardness of bamboo culms which suggests some parts of the plant possess a higher fiber density than others. Strength of the bamboo culms seems to differ between the top and bottom sections of the culm. This strength is dependent on density and diameter of the fibers, as well as the thickness and moisture content of the cell walls (Low and Che, 2006). Bamboo fibers are arranged in an alternating pattern of broad and narrow layers that have variations in the way the fibers are oriented (Jain *et al.*, 1992). This pattern of arrangement gives bamboo its high tensile strength and is not present in the fibers of normal wood (Jain et al., 1992).

DIET OF *PARANTHROPUS*

Paranthropus is a genus of robust hominins (robust referring to the large jaw and tooth size relative to other hominins) which currently contains the species *Paranthropus aethopicus, Paranthropus robustus,* and *Paranthropus boisei* (Wood and Constantino, 2007). Fossil evidence of *P. boisei* and *P. aethopicus* have been found in eight sites in East Africa and are dated to around 2.6 (Constantino and Wood, 2007) to 1.34 (Dominguez-Rodrigo *et al.*, 2013) million years ago. Most evidence of P. robustus comes from the south African sites of Swartkrans, Kromdraai, and Drimolen and has been dated to around 2 to 1 million years old. Morphological characters that are shared among these species include wide, flared zygomatic arches, ectocranial crests, and large postcanine teeth (Wood and Constantino, 2007). While some claim the morphology of *P. robustus* arose independently so it should not belong to the same genus (see Constantino and Wood, 2007), this paper will disregard the question of whether or not homoplasy is the cause of their similar morphological characteristics.

11

Biomechanically, it can be postdicted that *Paranthropus* had been a hard object feeder. Its robust cranial morphology, which includes features such as large zygomatic arches, a sagittal crest, and large molars covered in thick enamel, are indicative of strong bite forces and chewing power. Other organisms with these traits are known to consume hard foods such as nuts or seeds. Such organisms include sooty mangabeys (*Cercocebus atys*) which use their teeth to open hard nuts. Much like *Paranthropus*, sooty mangabeys have enlarged molars relative to their body size (Daegling et al, 2011). This adaptation seems to be well suited to hard object feeding because the large (and sometimes thickly enameled) molars can better withstand the high and often concentrated stresses placed on the teeth by hard food objects (Lucas *et al*, 2008). Although *P. robustus* seems to fit the prediction that the enhanced masticatory systems of its genus were used to eat hard foods as a fallback source of nutrition, its higher microwear complexity patterns indicate a varied diet consisting of tough foods as well as hard. These patterns are most similar to primates who rely on hard foods as sources of fallback nutrition (Ungar and Sponheimer, 2011).

P. boisei has microwear patterns which suggest it primarily consumed neither tough more than hard foods (Ungar, 2008). Confounding stable isotope results indicate *P. boisei* had a diet consisting of 75-80% C4 material (van der Merwe *et al.*, 2008; Cerling *et al.*, 2011; Ungar and Sponheimer, 2011). Plants using the C4 pathway are usually tough grasses or sedges and are not traditionally found in large quantities as part of the diet of extant primates. Chimpanzees, even those living in environments where C4 plants are plentiful, do not consume significant portions of C4. The almost exclusive consumption of either C4 grasses or organisms which feed upon those grasses is unique among hominins and is contrary to the diet inferred by *P. boisei*'s robust jaws and teeth. This behavior of eating a mostly C4 diet is similar to grass-eating warthogs, hippos, and zebras (Ungar and Sponheimer, 2011). Although there can exist a high variation of carbon isotope composition between taxa, there is no overlap in composition in *P. robustus* and *P. boisei*. The puzzling diet of *P. boisei* contrasted with the more expected diet of *P. robustus* could be evidence that the two species are not as closely related as their morphology might suggest (Wood, 1988).

Perhaps geography plays a significant role in the carbon isotope composition of organisms. Ungar and Sponheimer (2011) have discovered there is less variation in isotope compositions of the East African *P. boisei* than there is in the southern *P. robustus*. The same pattern was also true for microwear complexity. This may be because C_4 foods were/are more readily available in East Africa than southern Africa, although this would not explain why chimpanzees and other modern apes will ignore C_4 foods even when they are in high abundance. The geographic separation between *P. robustus* and *P. boisei* may provide evidence for their similar morphology taking on a new function. The robust jaws and teeth of *Paranthropus* may serve *P. robustus* by allowing it to consume hard nuts or seeds. This would fit the expected diet inferred by its tooth morphology. In contrast, *P. boisei* may have used the same adaptations for repetitious chewing of tough grasses or sedges. The low nutrient quality of these foods could have forced *P. boisei* to chew large quantities of these tough materials to meet their nutritional needs. This theory is weakened by the fact that living primates that exploit tough foods for fallback nutrition have sharp shearing crests which are used to slice through the tough, fibrous material (Lucas et al., 2008). *Paranthropus boisei* lacks shearing crests on its teeth and instead have large, flat molars which are better suiting to crushing and grinding hard materials. These teeth would have made eating tough sedges difficult, but perhaps *P. boisei* found a way to work with what it had. *P. boisei* may have been eating tough grasses or sedges in a way that no modern analog exists from which to draw comparisons. Chewing may have been inefficient, but perhaps the strong muscles of mastication compensated for this deficiency and allowed for extended periods of chewing. It should be noted that although the giant panda also possesses bunodont teeth with relatively large molar grinding areas (Sacco and Van Valkenburgh, 2004), it has no trouble consuming grasses in the form of bamboo.

A simple explanation for why hard foods do not appear in the microwear of *Paranthropus boisei* could be that none of the specimens on which microwear was examined contained evidence of hard object feeding because those specimens are not representative of *P. boisei* as a whole. Dental microwear only shows what the organism was eating in the short time before its death, so the specimens that were sampled may have died during a seasonal shortage of their preferred food. However, this explanation cannot account for the high degree of similarity among the samples tested as well as their varied temporal separation. It is unlikely the lack of hard object feeding evidence is due to sampling bias (Ungar *et al.*, 2008).

If in fact *Paranthropus boisei* ate sedges which grew near the water, its distribution may have been tied to these water sources. Given the low complexity of its diet, it stands to reason it could not live apart from its main food source for long. Perhaps its extinction was caused in part by its inability to travel away from water sources. Despite the efficiency of bipedal locomotion, *P. boisei* may have been unable to follow its primary source of food from one water source to the next. A lack of dietary diversity may have been the reason for the decline and ultimate disappearance of this hominin. A modern parallel for this explanation comes in the form of the giant panda. Giant pandas are dietary specialists which depend on bamboo for survival. Decreasing access to bamboo is a cited reason for the decline of the giant panda. If *Paranthropus boisei* was as highly specialized as the giant panda for eating low nutrient foods, then a lack of access to their main food source would be disastrous to their survival.

CONVERGENT MORPHOLOGY

If it is to be believed that the homoplasies between *Paranthropus* and giant pandas can indicate dietary similarity, evidence should be presented that links morphology to a food source. The robust skull and large molars of both the giant panda and the red panda (*Ailurus fulgens*) make them both suited to producing high bite forces which could explain their diet of bamboo (Christiansen and Wroe, 2007; Davis, 1964). Red pandas are not closely related to giant pandas, but although they are in different families, they both occupy a similar ecological niche (Pradhan *et al.*, 2001). A study on homoplasy in carnivore skulls has concluded that skull shape is correlated to feed behavior (Figueirido *et al.*, 2010). The researchers claim that the carnivores in the study that tend toward an herbivorous diet (which include both giant and red pandas) have shared traits in their craniodental anatomy. These traits include anteriorly positioned zygomatic arches, deep and short neurocrania, shortened premolars, and enlarged molar tooth rows. These characters are all positive indicators of an ability to generate high bite forces (Christiansen, 2007; Christiansen and Wroe, 2007).

SIGNIFICANCE

Learning more about the ecomorphology of the giant panda could allow further insight in the dietary history of the extinct hominin genus *Paranthropus*. The unique skull morphology of the giant panda is thought to be an adaptation to the mastication of bamboo (Christiansen, 2007). *Paranthropus* shares many cranial features of the giant panda as noted by both Davis (1964) and Du Brul (1977). Not only does the skull morphology of both creatures look similar, but both skulls are uniquely specialized among members of their own group. Both skulls are shorter and wider than the skulls of closely related species and both have deep and broad mandibles that contain large, bunodont, postcanine teeth.

Because of the specializations of both the giant panda and *Paranthropus*, it is reasonable to speculate both organisms derived their adaptations through similar means. Perhaps the food source of *Paranthropus* had qualities similar to those of bamboo. Both *Paranthropus* and the giant panda appear to be heavily specialized to chew and process food material. Studying the dietary specialization of the giant panda could be the key to discovering the selective pressures which drove the cranial adaptations and possible specialization of *Paranthropus*.

This research could be a starting point to further research into the dietary history of *Paranthropus*. If the mechanical properties of a variety of bamboo species are known, research which would look for specific foods available to *Paranthropus* which have similar mechanical properties to bamboo could reveal much about *Paranthropus's* diet. If bamboo has toughness similar to many grasses, this could suggest *Paranthropus* enjoyed a highly fibrous diet. Although the dental morphology of the teeth of *Paranthropus* argues against a diet of grasses, underground storage organs (USOs) such as tubers, seeds, roots, and rhizome are not unreasonable (Wood and Constantino, 2007; Dominy, *et. al*. 2008; van der Merwe *et al*., 2008). The diet of *Paranthropus* is still largely unknown, but perhaps this research will aid further attempts to uncover more about this specialized hominin.

As well as bamboo being the primary food source of the giant panda, bamboo is also of great interest as a building material. In the construction industry, bamboo is often used as scaffolding because of its low cost, easy access, and general stability (Low and Che, 2006). Bamboo is comparable to materials such as low carbon steel and glass reinforced plastics because of its high elastic modulus and compressive strength (Low and Che, 2006). The fibrous makeup of bamboo gives it strength and makes it a cheap and environmentally safe alternative to many conventional materials. Because this research will be sampling a variety of bamboo species, perhaps one species of bamboo will outperform others and would be a more suitable building material. Both young and adult bamboo samples will be tested and the results will provide information about the strength of aging bamboo. These data could be useful in determining at what age bamboo becomes most suitable for building material.

CHAPTER 2

METHODS AND RESULTS

MATERIALS AND METHODS

BAMBOO SELECTION

The bamboo species which were selected for testing came from two categories. First, we selected bamboo which giant pandas are known to eat. This bamboo should yield the best data in regards to stress placed on the masticatory systems of giant pandas. Bamboo species which occur naturally in the giant panda's habitat were selected as the second category. There is no consensus in the scientific literature on all species of bamboo pandas are known to eat, so a wide selection of bamboo found in their habitat should provide sufficient samples of bamboo pandas could be eating. This category also aims to discover if there are any differences in the properties of the bamboos that are favored by giant pandas and bamboos that are not known to be ingested.

Four bamboo species were selected for this study. These bamboos are *Pseudosasa japonica, Phyllostachys nigra, Phyllostachys bissetii,* and *Phyllostachys dulcis.* The first three species were selected based on the results of a preference study conducted on giant pandas which aimed to determine which species of bamboo giant pandas are most likely to select in the wild (Tarou *et al.*, 2005). *Phyllostachys dulcis* was not included in the preference study, but is both abundantly found in China and grown for food because of its sweet taste.

The selected bamboo species were purchased and shipped from MidAtlantic Bamboo, a bamboo nursery, and were kept well nourished in a greenhouse until the time of testing. The bamboos received one hour of light watering per day provided by an automatic irrigation system. The plants' leaves were also occasionally misted with water to ensure they did not dry out. Leaves were tested within one hour of being removed from the parent plant to ensure their condition most closely reflected leaves being eaten from a wild bamboo.

Both young and adult bamboos were tested. Giant pandas seem to prefer younger shoots (thicker young shoots are given preference) to older bamboo (Wei *et al.*, 1999) so differences in the mechanical properties of young and adult bamboos were recorded. The parts of the bamboo giant pandas prefer to eat are variable depending on the season, so property data were collected using both the leaves and stems of the bamboo plants.

Figure 1. Comparison of young and adult P. nigra. *Young* P. nigra *(left) is approximately 3 to 4 months old while the adult (right) is 6 to 8 months old. The other purchased bamboo (not pictured) is similar in size to* P. nigra.

PROPERTIES TESTS

The tests were performed using an FLS-1 universal testing machine supplied by Lucas Scientific (see Figure 4). Toughness values were acquired through the use of a scissors test for both the bamboo stems and leaves. Scissor testing is an effective technique when sampling homogenous materials in sheet form (Darvell *et al.*, 1996). Scissors tests are performed by cutting the leaves and stems with scissors by slowly applying force with the tester. Turning the handle smoothly lowers the crossbar onto the handle of the scissors which then cuts through the specimen (see Figure 2). To make the stems suitable for this test, individual fibers were stripped away from the bamboo stem and anchored to a paper towel so the scissor could smoothly cut through the paper, cutting the fibers along with it. The FLS-1 software requires an empty pass (closing the scissors without cutting the plant material) be made before the actual test to measure the amount of friction caused by the scissor blades passing one another. The test removes the background friction from the actual test to accurately measure the toughness of the sample material only. In the case of the bamboo stems, the paper towel was cut on the empty pass to remove its toughness from the result of the actual test.

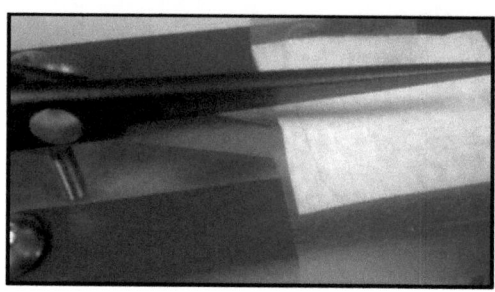

Figure 2. Scissors Test. Bamboo fibers are affixed to a paper towel when cutting to prevent the fiber from moving during the cut. Both leaves and paper towels are anchored to the stage by tape to prevent movement. Caution must be taken during taping to ensure no tape is in the path of the scissor blades.

Hardness data of stems only were obtained through Vickers indentation. Vickers indentation is a hardness test in which a pyramidal indenter is used to apply a known, steady force to a material until the material becomes plastically deformed. Bamboo specimens were prepared by cutting the stems so that they could lay flat against the anvil (see Figure 2). The indenter tip was smoothly lowered on the bamboo sample so that it penetrated the stem no further than 1mm. The indenter tip was then smoothly retracted from the bamboo to measure the degree in which the sample plastically deforms, which is then used to calculate hardness.

Young's modulus is defined as the ratio of stress over strain and is the relationship between the force on the object and the displacement caused by said force (Lucas *et al.*, 2000). The force is measured by the load cell and the change in the length (displacement) of the specimen is measured by a displacement calculation on the tester. Young's modulus is roughly equivalent to the stiffness of a material. These data were found by measuring the area of a bamboo specimen and then applying force. Data were collected using a 4-point bending test (see Figure 2). 4-point bending tests apply pressure to the specimen in four points, two from above and two from below. Pressure is applied to the specimen until elastic deformation occurs. The specimen does not need to fail under the pressure as stiffness is a measure of elasticity.

 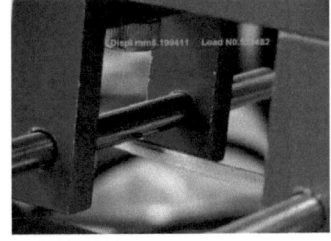

A. B.

Figure 3. Vicker's Indentation and 4 Point Bending. A (left) demonstrates Vicker's indentation. The indenter tip is smoothly lowered onto the bamboo specimen to plastically deform the structure. The specimen is cut flat so that it rests squarely on the base of the tester. B (right). Figure B demonstrates 4-point bending. Two rods apply pressure from above while the specimen is supported from below by two more rods (not pictured). The force bends the specimen and Young's modulus is calculated by measuring the dimensions of the sample and the amount of force applied.

Data collected by the load cell and displacement counter are sent to a personal computer which displays a real time graph of the forces acting on the bamboo sample and the displacement caused by loading.

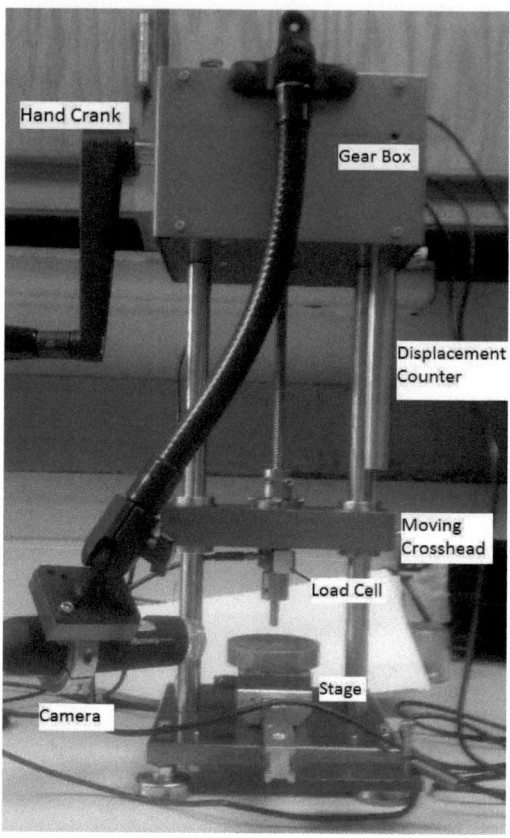

Figure 4. The FSL-1 Portable Testing Machine. Force is generated by turning the hand crank which lowers the moving crosshead. The displacement counter records how far the crosshead travels and the force placed on the specimen is picked up by the load cell and set to a personal computer. The camera records pictures and video of the specimen during tests.

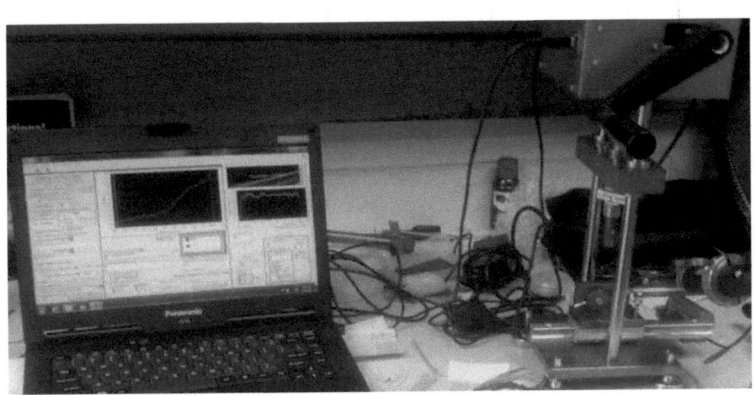

Figure 5. FLS-1 Tester Along with Complimentary Software. Information from the tester is sent to the computer and is graphically displayed in the program. Measurements are inputted to the computer and are used to calculate various mechanical properties. Tester is currently equipped for scissors tests.

Data collected from these tests was compiled and analyzed using the statistical software JMP. JMP was used to perform t-tests and ANOVAs to analyze collected data and graphs were created using JMP to represent said data.

SPECIMEN PREPARATION

Before certain tests could begin, bamboo specimens must be cut to fit the test. For both hardness and Young's modulus tests, bamboo specimens were cut into flat, rectangular pieces. This was achieved by cutting a section in between nodes and bisecting the section lengthwise into two semicircular halves. Each half was bisected again and depending on the size of the specimen, once more after that until the culm of the bamboo rested flat on the surface of the tester. For fiber collection, individual bamboo fibers were peeled from the untested specimens that had been bisected for hardness and modulus tests. One fiber constituted the smallest strand of bamboo that was able to be removed without the aid of a microscope.

COMPARISON TO WILD BAMBOO

In order to ensure the bamboos tested are structurally similar to bamboos found in the wild, data from the purchased bamboo samples were compared to data collected at Foping Nature Reserve in central China in 2006 by Dr. Paul Constantino. Foping is home to the highest concentration of wild giant pandas and data from bamboo samples collected there accurately represent the properties of wild bamboo. In addition to comparing structural similarity, data collected from this study are combined with the data collected by Dr. Paul Constantino to increase the sample size of bamboo tested as well as provide data on how bamboo mechanical properties differ among species.

RESULTS

LEAF TOUGHNESS

The analysis of the leaves of both young and adult bamboo species reveals that there are differences in leaf toughness among species (Figure 6). Both young and adult bamboo are grouped together to increase sample sizes, particularly in the cases where fewer specimens where tested. *P. japonica* and *P. nigra* appear to have the toughest leaves and both have a toughness of over 2000 Jm^{-2}. *P. dulcis* and *P. bissetii* are not as tough with toughness values of around 1200 Jm^{-2} and 1400 Jm^{-2}, respectively. *P. dulcis* and *P. bissetii* do not appear to be significantly different from each other and neither do *P. japonica* and *P. nigra*. The difference between *P. japonica* and *P. dulcis* is significant ($p < .0001$) as is the difference between *P. japonica* and *P. bissetii* ($p < .0001$). *P. nigra* is significantly different from *P. dulcis* and *P. bissetii* (both $p < 0.0001$).

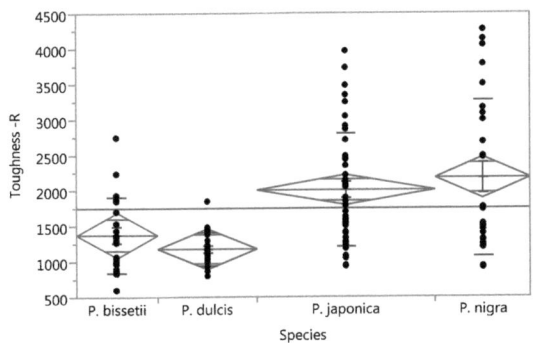

Figure 6. Leaf Toughness of Young and Adult Bamboo. Leaf toughness results indicate P. japonica *and* P. nigra *are tougher than* P. dulcis *and* P. bissetii *and are of similar toughness. Toughness of bamboo leaves is variable among species (p < 0.0001). Sample sizes are as follows:* P. bissetii *(22),* P. dulcis *(28),* P. japonica *(51),* P. nigra *(26).*

STEM TOUGHNESS

The fibers of *P. nigra* are significantly tougher than the fibers of the other species tested ($p < 0.0001$). The toughness of *P. nigra* stems is over 9600 Jm^{-2}, *P. bissetii* has a stem toughness of around 6700 Jm^{-2}, and *P. dulcis* and *P. japonica* both have stem toughness values of around 5600 Jm^{-2}. Other than *P. nigra*, there appear to be no significant differences in the stem toughness of the species (Figure 7).

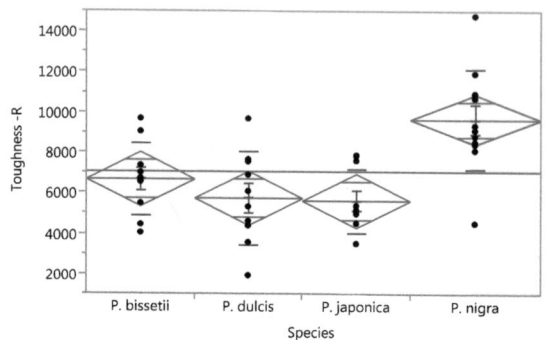

Figure 7. Stem Fiber Toughness of Young and Adult Bamboo. P. nigra *has the toughest stems off all tested bamboo and is significantly tougher than the other species (p < 0.0001). The other species are not significantly different from one another. Sample sizes are as follows:* P. bis*setii (10),* P. dulcis *(10),* P. japonica *(10),* P. nigra *(12).*

STEM HARDNESS

P. bissetii appears to be the hardest bamboo with an average hardness of about 14 megapascals (MPa). Although the difference between *P. bissetii* and *P. japonica* are not significantly different, *P. bissetii* is harder than both *P. dulcis* (p < 0.0143) and *P. nigra* (p < 0.003). *P. japonica* is harder than *P. nigra* (p < 0.0168) but is not significantly different from any of the other tested species. *P. dulcis* and *P. nigra* appear to be the least hard of the bamboo tested.

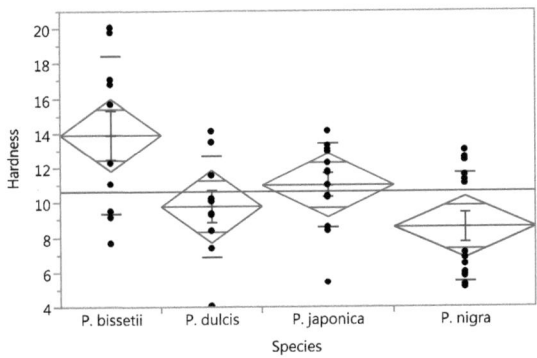

Figure 8. Hardness of Young and Adult Bamboo. P. bissetii *is the hardest bamboo tested, with* P. nigra *being the least hard. Both* P. bissetii *and* P. japonica *are harder than* P. nigra. *No difference was found between* P. nigra *and* P. dulcis. *Sample sizes are as follows:* P. bissetii *(10),* P. dulcis *(10),* P. japonica *(12),* P. nigra *(12).*

YOUNG'S MODULUS (STEM)

No significant differences were found in the Young's modulus of the tested bamboo species. Each species was found to have a Young's modulus of slightly over 1.1 gigapascals (GPa). *P. bissetii* has the highest mean modulus with 1.5 GPa, but this value is not significantly higher than the other bamboo. This pattern is partially repeated when analyzing only adult bamboo, however both *P. japonica* and *P. bissetii* have a higher average modulus than *P. nigra* (p < 0.026 and p < 0.0027 respectively). Strangely, *P. nigra* has a higher modulus than *P. bissetii* when examining only young bamboo stems (p < 0.0436). No other elastic modulus differences are found in young plants.

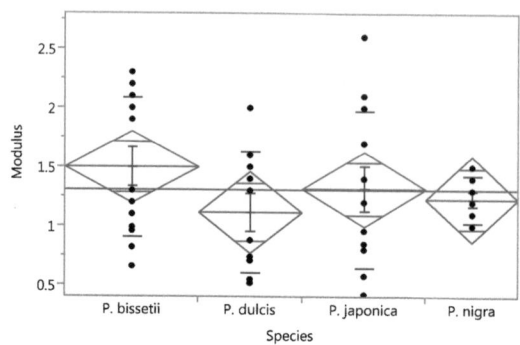

Figure 9. Young's Modulus of Young and Adult Bamboo. No significant differences in modulus were found for any of the bamboo species tested. Sample sizes are as follows: P. bissetii *(12),* P. dulcis *(10),* P. japonica *(12),* P. nigra *(9).*

Table 1. Comparison of Bamboo Mechanical Properties. Toughness is measured in units of Jm^{-2}, *hardness is in megapascals, and Young's modulus is measured in gigapascals.*

Comparison of Bamboo Mechanical Properties					
Property	*P. japonica*	*P. bissetii*	*P. dulcis*	*P. nigra*	Comparison
Toughness (stems)	5625 ± 1570	6696 ± 1791	5758 ± 2280	9662 ± 2450	*P. nigra* has the toughest stems, all other species are not significantly different
Toughness (leaves)	2015 ± 795	1383 ± 526	1186 ± 234	2186 ± 1098	*P. nigra* and *P. japonica* have the

27

					toughest leaves, *P. bissetii* and *P. dulcis* are not significantly different
Hardness	11.0 ± 2.4	13.9 ± 4.5	9.8 ± 2.9	8.6 ± 3.1	*P. dulcis* and *P. nigra* are least hard, *P. bissetii* is hardest but not significantly different from *P. japonica*
Young's Modulus	1.3 ± 0.67	1.5 ± 0.59	1.1 ± 0.51	1.2 ± 0.20	No significant differences in modulus found

YOUNG VS ADULT BAMBOO

The trends in leaf toughness are continued when young and adult bamboos are analyzed separately. In young plants, *P. japonica* and *P. nigra* are still the toughest leaves with *P. dulcis* and *P. bissetii* being of comparable toughness. *P. japonica* is tougher than both *P. bissetii* ($p < 0.0042$) and *P. dulcis* ($p < 0.0001$), but is less tough than *P. nigra* ($p < 0.0177$). *P. nigra* seems to have leaves that are tougher in young plants than in adults. The young leaves have an average toughness of over 2800 Jm^{-2}

while older leaves have a toughness of only 2000 Jm^{-2}. This is a significant decrease in toughness as the plant ages ($p < 0.0193$).

For adult leaves, *P. japonica* and *P. nigra* are of similar toughness and are still tougher than the other bamboo species. Adult *P. japonica* is tougher than *P. bissetii* ($p < 0.0005$) and *P. dulcis* ($p < 0.0001$) and adult *P. nigra* is tougher than *P. bissetii* ($p < 0.0241$) and *P. dulcis* ($p < 0.0026$). Once again, *P. bissetii* and *P. dulcis* are the least tough and have comparable toughness values of around 1200-1400 Jm^{-2}.

When young bamboo fibers are examined, *P. japonica* appears to be the least tough (6000 Jm^{-2}) while *P. nigra* remains the bamboo with the toughest stems (over 9000 Jm^{-2}). Young *P. nigra* is tougher than *P. japonica* ($p < 0.0125$) but is not significantly tougher than the other bamboos tested. Young *P. bissetii* stems are also tougher than *P. japonica* ($p < 0.045$). All other young stems are similar in toughness. In adult bamboo, *P. nigra* also has the toughest stems. It is significantly tougher than *P. japonica* ($p < 0.0068$), *P. bissetii* ($p < 0.0104$), and *P. dulcis* ($p < 0.0006$). The other species that were tested have adult stem toughness values that are comparable to each other.

In adult bamboo, *P. japonica* and *P. bissetii* seem to be most hard. There was no difference found between *P. japonica* and *P. bissetii*, however, both *P. japonica* and *P. bissetii* are harder than *P. dulcis* ($p < 0.0313$ and $p < 0.0186$ respectively). No significant differences were found between adult *P. nigra* and any other species. In young plants, *P. bissetii* is the hardest species that was tested. It has a higher hardness value than *P. japonica* ($p < 0.0485$) and *P. nigra* ($p < 0.0058$), but is not significantly different than *P. dulcis*. *P. japonica* and *P. dulcis* both seem to be harder than *P. nigra* ($p < 0.0067$ and $p < 0.0044$), making *P. nigra* the least hard of the young bamboo. Adult bamboo ranges from 9-11.5 megapascals and all species are similar in average hardness. Young bamboo is much more variable and ranges from 7 MPa (*P. nigra*) to 16 MPa (*P. bissetii*).

There are differences between young and adult bamboo. In *P. japonica*, leaf toughness and stem hardness are equivalent between young and adult, but adult stems are tougher and have a higher elastic modulus than young stems ($p < 0.0331$ and $p < 0.0083$). *P. bissetii* has adult plants that have a higher modulus than young plants ($p <$

0.0006), but there are no other significant differences between young and adult. In *P. dulcis*, young bamboo is actually harder than adult bamboo (p < 0.0059), but there are no other differences to report. Young *P. nigra* has leaves that are tougher than adult bamboo (p < 0.0193). No other differences between young and adult *P. nigra* were found.

Table 2. Comparison of Adult Bamboo. Units of toughness are J/m^{-2}, units of hardness are MPa, and units of Young's modulus are GPa.

Comparison of Adult Bamboo					
Property	*P. japonica*	*P. bissetii*	*P. dulcis*	*P. nigra*	Comparison
Toughness (Stems)	6561 ± 1497	6723 ± 1619	5075 ± 1363	7411 ± 2903	*P. nigra* stems are tougher than all other species. No significant differences found among *P. japonica*, *P. bissetii*, or *P. dulcis*.
Toughness (Leaves)	2022 ± 645	1302 ± 594	1102 ± 182	1434 ± 412	*P. japonica* leaves are tougher than *P. bissetii* and *P. dulcis*. *P. nigra* leaves are tougher than *P. bissetii* and *P. dulcis*. No significant differences in *P. japonica* and *P. nigra* and between *P. bissetii* and *P. dulcis*.
Hardness	11.5 ± 2.2	11.6 ± 2.4	7.7 ± 1.9	9.0 ± 3.4	*P. japonica* and *P. bissetii* are both harder than *P. dulcis*. No

					other significant differences found.
Young's Modulus	1.86 ± 0.5	1.84 ± 0.47	1.26 ± 0.29	1.12 ± 0.11	Both *P. japonica* and *P. bissetii* have a higher modulus than *P. nigra*. No other significant differences found.

Table 3. Comparison of Young Bamboo. Units of toughness are J/m^{-2}, units of hardness are MPa, and units of Young's modulus are GPa.

Comparison of Young Bamboo					
Property	*P. japonica*	*P. bissetii*	*P. dulcis*	*P. nigra*	Comparison
Toughness (Stems)	4690 ± 664	6670 ± 1776	6442 ± 2563	9259 ± 2685	*P. nigra* and *P. bissetii* are both tougher than *P. japonica*. No other significant differences found.
Toughness (Leaves)	2019 ± 838	1479 ± 375	1259 ± 242	2824 ± 740	*P. nigra* leaves are significantly tougher than other species. *P. japonica* leaves are tougher than *P. bissetii* and *P. nigra*.

					P. bissetii and *P. dulcis* leaves are not significantly different.
Hardness	11.4 ± 0.67	16.3 ± 4.5	12.4 ± 1.5	7.2 ± 2.4	All species are significantly harder than *P. nigra*. *P. bissetii* is harder than *P. japonica*. Both *P. bissetii* and *P. japonica* are not significantly different to *P. dulcis*.
Young's Modulus	0.93 ± 0.39	0.96 ± 0.22	0.98 ± 0.36	1.28 ± 0.23	*P. nigra* has a higher Young's modulus than *P. bissetii*. No other differences were found.

BONFERRONI CORRECTION

To guard against false positives when performing Student's t-tests on these data, a Bonferroni correction was made that modifies the p value that indicates significant difference. After the correction, the new p value is 0.00057 which was calculated by dividing the original significance marker of 0.05 by the number of t-tests performed (88). With the adjusted p value in place, the significance of the data is altered and is reported below.

TOUGHNESS OF BAMBOO LEAVES

All Ages

P. japonica leaves are tougher than *P. bissetii* ($p < 0.0002$) and *P. dulcis* ($p < 0.0001$), but is not significantly different from *P. nigra*. *P. nigra* is tougher than *P. dulcis* ($p < 0.0001$), but is not significantly different from any other tested species. No other significant differences are found after adjustment.

Young Leaves

P. japonica has tougher leaves than *P. dulcis* ($p < 0.0001$), but no other differences are detected. No other significant differences are found in young leaves after adjustment.

Adult Leaves

P. japonica has tougher leaves than *P. bissetii* ($p < 0.0005$) and *P. dulcis* ($p < 0.0001$), but is similar in toughness to *P. nigra*. No other differences found after adjustment.

TOUGHNESS OF BAMBOO STEMS

All Ages

The stems of *P. nigra* are tougher than *P. japonica* ($p < 0.0002$), but no other differences are found between any other tested species.

Young Stems

There are no significant differences to report in young bamboo stems after adjustment.

Adult Stems

No significant differences are found after adjustment.

HARDNESS OF BAMBOO STEMS

After adjustment, there are no significant differences to report among any of the species or ages tested.

YOUNG'S MODULUS OF STEMS

There are no significant differences to report among any ages or species after adjustment.

YOUNG VS ADULT BAMBOO

There are no significant differences to report among species regardless of which property was tested. After the Bonferroni correction, all young and adult bamboo species are reported to be similar in all properties.

Table 4. Significance of bamboo comparisons with and without Bonferroni correction. This table summarizes the results of each t-tests performed to determine the significance of each comparison of material properties. Approximate values are given in order given in the previous column. Significance is determined by having a p value less than 0.05 and significance after Bonferroni's correction is determined at a p value less than 0.00057.

	T-test	Values	Significance	Bonferroni Correction
	P. nigra vs *P. dulcis*	$9000 - 5700$ Jm^{-2}	Significant (p < .001)	Not Significant
	P. bissetii vs *P. dulcis*	$6500 - 5700$ Jm^{-2}	Not Significant	Not Significant
Toughness of Stems (All Ages)	*P. japonica* vs *P. dulcis*	$5200 - 5700$ Jm^{-2}	Not Significant	Not Significant
	P. bissetii vs *P. nigra*	$6500 - 9000$ Jm^{-2}	Significant (p <.0041)	Not Significant
	P. bissetii vs *P japonica*	$6500 - 5200$ Jm^{-2}	Not Significant	Not Significant
	P. nigra vs *P. japonica*	$9000 - 5200$ Jm^{-2}	Significant (p < .0002)	Significant
	P. japonica vs *P. nigra*	$2000 - 2200$ Jm^{-2}	Not Significant	Not Significant
	P. japonica vs *P. bissetii*	$2000 - 1350$ Jm^{-2}	Significant (p < .0002)	Significant
Leaf Toughness (All Ages)	*P. japonica* vs *P. dulcis*	$2000 - 1300$ Jm^{-2}	Significant (p < .0001)	Significant
	P. bissetii vs *P. dulcis*	$1350 - 1300$ Jm^{-2}	Not Significant	Not Significant
	P. bissetii vs *P. nigra*	$1350 - 2200$ Jm^{-2}	Significant (p < .0021)	Not Significant
	P. nigra vs *P. dulcis*	$2200 - 1300$ Jm^{-2}	Significant (p < .0001)	Significant

	P. japonica vs *P. bissetii*	11 – 13.9 MPa	Not Significant	Not Significant
	P. japonica vs *P. dulcis*	11 – 9.8 MPa	Not Significant	Not Significant
Stem Hardness (All Ages)	*P. japonica* vs *P. nigra*	11 – 8.6 MPa	Significant (p < .0168)	Not Significant
	P. bissetii vs *P. dulcis*	13.9 – 9.8 MPa	Significant (p < .0143)	Not Significant
	P. bissetii vs *P. nigra*	13.9 – 8.6 MPa	Significant (p < .003)	Not Significant
	P. dulcis vs *P. nigra*	9.6 – 8.6 MPa	Not Significant	Not Significant
	P. japonica vs *P. bissetii*	1.3 – 1.5 GPa	Not Significant	Not Significant
	P. japonica vs *P. dulcis*	1.3 – 1.1 GPa	Not Significant	Not Significant
Young's Modulus of Stems (All Ages)	*P. japonica* vs *P. nigra*	1.3 – 1.2 GPa	Not Significant	Not Significant
	P. bissetii vs *P. dulcis*	1.5 – 1.1 GPa	Not Significant	Not Significant
	P. bissetii vs *P. nigra*	1.5 – 1.2 GPa	Not Significant	Not Significant
	P. dulcis vs *P. nigra*	1.1 – 1.2 GPa	Not Significant	Not Significant
	P. japonica vs *P. bissetii*	6500 – 6700 Jm^{-2}	Not Significant	Not Significant
Stem Toughness (Adult)	*P. japonica* vs *P. dulcis*	6500 – 5200 Jm^{-2}	Not Significant	Not Significant
	P. japonica vs *P. nigra*	6500 – 10000 Jm^{-2}	Significant (p < .0068)	Not Significant
	P. bissetii vs *P. dulcis*	6700 – 5200 Jm^{-2}	Not Significant	Not Significant

	P. bissetii vs *P. nigra*	6700 – 10000 Jm^{-2}	Significant (p < .0104)	Not Significant
	P dulcis vs *P. nigra*	5200 – 10000 Jm^{-2}	Significant (p < .0006)	Not Significant
Leaf Toughness (Adult)	*P. japonica* vs *P. bissetii*	2000 – 1250 Jm^{-2}	Significant (p < .0005)	Significant
	P. japonica vs *P. dulcis*	2000 – 1200 Jm^{-2}	Significant (p < .0001)	Significant
	P. japonica vs *P. nigra*	2000 – 1950 Jm^{-2}	Not Significant	Not Significant
	P. bissetii vs *P. dulcis*	1250 – 1200 Jm^{-2}	Not Significant	Not Significant
	P. bissetii vs *P. nigra*	1250 – 1950 Jm^{-2}	Significant (p < .0241)	Not Significant
	P. dulcis vs *P. nigra*	1200 – 1950 Jm^{-2}	Significant (p < .0026)	Not Significant
Stem Hardness (Adult)	*P. japonica* vs *P. bissetii*	11 – 12 MPa	Not Significant	Not Significant
	P. japonica vs *P. dulcis*	11 – 8 MPa	Significant (p < .0313)	Not Significant
	P. japonica vs *P. nigra*	11 – 9 MPa	Not Significant	Not Significant
	P. bissetii vs *P. dulcis*	12 – 8 MPa	Significant (p < .0186)	Not Significant
	P. bissetii vs *P. nigra*	12 – 9 MPa	Not Significant	Not Significant
	P. dulcis vs *P. nigra*	8 – 9 MPa	Not Significant	Not Significant
Young's Modulus of Stems (Adult)	*P. japonica* vs *P. bissetii*	1.8 – 1.8 GPa	Not Significant	Not Significant
	P. japonica vs *P. dulcis*	1.8 – 1.3 GPa	Not Significant	Not Significant

	P. japonica vs *P. nigra*	1.8 – 1.2 GPa	Significant (p < .0216)	Not Significant
	P. bissetii vs *P. dulcis*	1.8 – 1.3 GPa	Not Significant	Not Significant
	P. bissetii vs *P. nigra*	1.8 – 1.2 GPa	Significant (p < .0027)	Not Significant
	P. dulcis vs *P. nigra*	1.3 – 1.2 GPa	Not Significant	Not Significant
Stem Toughness (Young)	*P. japonica* vs *P. bissetii*	4500 – 6500 Jm^{-2}	Significant (p < .045)	Not Significant
	P. japonica vs *P. dulcis*	4500 – 6400 Jm^{-2}	Not Significant	Not Significant
	P. japonica vs *P. nigra*	4500 – 8000 Jm^{-2}	Significant (p < .0125)	Not Significant
	P. bissetii vs *P. dulcis*	6500 – 6400 Jm^{-2}	Not Significant	Not Significant
	P. bissetii vs *P. nigra*	6500 – 8000 Jm^{-2}	Not Significant	Not Significant
	P. dulcis vs *P. nigra*	6400 – 8000 Jm^{-2}	Not Significant	Not Significant
Leaf Toughness (Young)	*P. japonica* vs *P. bissetii*	1900 – 1500 Jm^{-2}	Significant (p < .0042)	Not Significant
	P. japonica vs *P. dulcis*	1900 – 1300 Jm^{-2}	Significant (p < .0001)	Significant
	P. japonica vs *P. nigra*	1900 – 2650 Jm^{-2}	Significant (p < .0177)	Not Significant
	P. bissetii vs *P. dulcis*	1500 – 1300 Jm^{-2}	Not Significant	Not Significant
	P. bissetii vs *P. nigra*	1500 – 2650 Jm^{-2}	Significant (p < .0015)	Not Significant
	P. dulcis vs *P. nigra*	1300 – 2650 Jm^{-2}	Significant (p < .0008)	Not Significant

Stem Hardness (Young)	*P. japonica* vs *P. bissetii*	11.5 – 16 MPa	Significant (p < .0485)	Not Significant
	P. japonica vs *P. dulcis*	11.5 – 12 MPa	Not Significant	Not Significant
	P. japonica vs *P. nigra*	11.5 – 7 MPa	Significant (p < .0067)	Not Significant
	P. bissetii vs *P. dulcis*	16 – 12 MPa	Not Significant	Not Significant
	P. bissetii vs *P. nigra*	16 – 7 MPa	Significant (p < .0058)	Not Significant
	P. dulcis vs *P. nigra*	12 – 7 MPa	Significant (p < .0044)	Not Significant
Young's Modulus of Stems (Young)	*P. japonica* vs *P. bissetii*	0.9 – 0.9 GPa	Not Significant	Not Significant
	P. japonica vs *P. dulcis*	0.9 – 1 GPa	Not Significant	Not Significant
	P. japonica vs *P. nigra*	0.9 – 1.3 GPa	Not Significant	Not Significant
	P. bissetii vs *P. dulcis*	0.9 – 1 GPa	Not Significant	Not Significant
	P. bissetii vs *P. nigra*	0.9 – 1.3 GPa	Significant (p < .0436)	Not Significant
	P. dulcis vs *P. nigra*	1 – 1.3 GPa	Not Significant	Not Significant

Table 5. Summary of t-tests comparing young and adult bamboo. In the value column, approximate values for young bamboo are listed before adult. Significance is determined by having a p value less than 0.05 and significance after Bonferroni's correction is determined at a p value less than 0.00057.

Young vs Adult Bamboo				
Property	Species	Value	Significance	Bonferroni Correction
Leaf Toughness	*P. japonica*	$1900 - 2000$ Jm^{-2}	Not Significant	Not Significant
Stem Toughness	*P. japonica*	$4500 - 6500$ Jm^{-2}	Adult Tougher ($p < .0331$)	Not Significant
Hardness	*P. japonica*	$11.5 - 11$ MPa	Not Significant	Not Significant
Young's Modulus	*P. japonica*	$0.9 - 1.8$ GPa	Adult Higher Modulus ($p < .0083$)	Not Significant
Leaf Toughness	*P. bissetii*	$1500 - 1250$ Jm^{-2}	Not Significant	Not Significant
Stem Toughness	*P. bissetii*	$6500 - 6700$ Jm^{-2}	Not Significant	Not Significant
Hardness	*P. bissetii*	$16 - 12$ MPa	Not Significant	Not Significant
Young's Modulus	*P. bissetii*	$0.9 - 1.8$ MPa	Adult Higher Modulus ($p < .0006$)	Not Significant
Leaf Toughness	*P. dulcis*	$1300 - 1200$ Jm^{-2}	Not Significant	Not Significant
Stem Toughness	*P. dulcis*	$6400 - 6700$ Jm^{-2}	Not Significant	Not Significant
Hardness	*P. dulcis*	$12 - 8$ MPa	Young Harder ($p < .0059$)	Not Significant
Young's Modulus	*P. dulcis*	$1 - 1.3$ MPa	Not Significant	Not Significant

Leaf Toughness	*P. nigra*	$2650 - 1950$ Jm^{-2}	Young Tougher ($p < .0193$)	Not Significant
Stem Toughness	*P. nigra*	$8000 - 10000$ Jm^{-2}	Not Significant	Not Significant
Hardness	*P. nigra*	$7 - 9$ MPa	Not Significant	Not Significant
Young's Modulus	*P. nigra*	$1.3 - 1.2$ MPa	Not Significant	Not Significant

CHAPTER 3

DISCUSSION AND CONCLUSIONS

DISCUSSION

PATTERNS IN MATERIAL PROPERTIES OF BAMBOO

For the bamboo species tested, certain trends emerged. *P. nigra* is the toughest bamboo that was tested and is significantly tougher than the other bamboos in both leaves and stems. (Note that the initial significance and not the significance after Bonferroni's correction is being discussed here. Bonferroni's correction is not discussed as it is believed to have produced a number of false negatives which limit the conclusions able to be drawn.) *P. nigra* also happens to be the least hard of the bamboo tested, but is of similar hardness to *P. dulcis*. This pattern agrees with the idea that materials are either stress or displacement limited in their mechanical defenses (Lucas, 2004). As Lucas *et al.* (2000) describe in their paper on mechanical defenses to herbivory, stress limited defenses rely on a hard exterior to avoid fracture while displacement limited defenses focus on preventing the propagation of cracks that have already started. Bamboo seems to have a displacement limited defense as the stems are not very hard, but are tough. It has been proposed that different tooth sizes are more effective for eating foods that are stress or displacement limited (Lucas, 2004; Ungar and Lucas, 2010). According to a habitat appraisal study in Mount Shennongjia in Central China, *P. nigra* is an acceptable food source of the giant panda (Li and Denich, 2004). The tough bamboo material of *P. nigra* does not seem to dissuade giant pandas from selecting this bamboo over other species within its habitat. As Christiansen and Wroe (2007) have stated, the skull morphology of the giant panda allows it to generate high bite forces relative to its size. Figuerido *et al.*, (2010) have elaborated on this subject and claim the giant panda's morphology makes it more suited to an herbivorous diet than the other carnivorous ursines. These adaptations toward an herbivorous diet (enlarged molars, shortened skull, flared zygomatic arches) seem to allow for the consumption of tough materials when selecting a food source.

No significant differences in Young's modulus were detected among any of the bamboo species tested. Although some species have higher hardness values than others (*P. bissetii* being the hardest), this does not seem to have a significant impact on the overall stiffness of the material. Since rigid and stiff materials yield higher Young's modulus values than flexible ones (Smith and Walmsley, 1959), it should be expected that older and more rigid bamboo should have a higher modulus. Although no statistically significant modulus differences were found among species, the adult plants of some species have higher modulus values than the younger plants.

ADULT VS YOUNG BAMBOO

When comparing adult and young bamboo, few differences in material properties were found. In both *P. japonica* and *P. bissetii*, adult plants have a higher Young's modulus (although no difference in modulus between young and adult was found in *P. dulcis* and *P. nigra*.) This is not surprising as older bamboo has had more time to increase in lignin content, thus increasing the overall rigidity of the plant (Liese and Weiner, 1996). Lignin, along with cellulose and hemicellulose, are responsible for the rigid structure of bamboo culms and is also concentrated in bamboo leaves (Lin *et al.*, 2002). Lin *et al.* (2002) have also reported that the lignification process can continue after the plant has reached maturity and finishes growing. Unexpectedly, young *P. dulcis* is found to be harder than the adult and young *P. nigra* leaves are tougher than adults of the same species. Both results are surprising because the lignin content of adult bamboo should be greater, or at least equal to, that of young bamboo (Liese and Weiner, 1996). The cell walls of bamboo continue to thicken as the plant ages, also contributing to overall toughness (Alvin and Murphy, 1988; Lucas *et al*, 2000).

They most likely explanation for these unexpected differences in toughness and hardness is that the plants were not significantly different in age. The young bamboo which was tested in this study had an age of 3-4 months while the adult bamboo was about 6-8 months old. Perhaps this age difference is not significant enough for any real differences in mechanical properties to be revealed.

COMPARISON OF BAMBOO TO OTHER MATERIALS

To illustrate how the properties of bamboo found in this study apply to panda feeding, comparisons should be made with other materials. The following data were collected by Dr. Paul Constantino at Foping Nature Reserve in Central China (unpublished results). Two species of bamboo were tested for properties of hardness, Young's modulus, and leaf toughness (Table 1).

Table 1. Material Properties of Bamboo in Foping Nature Reserve. Both young (1 year) and adult (2 years) bamboo was collected for testing. The bamboo species tested were Fargesia qinlingensis and Bashaina fargesii, both fed on by giant pandas in this habitat.

Species	Age	Property		
		Hardness	Toughness (Leaf)	Modulus
Fargesia	Young	1.1 MPa	508 Jm^{-2}	3.4 GPa
qinlingensis	Adult	2.82 MPa	366 Jm^{-2}	7.8 GPa
Bashaina	Young	1.76 MPa	864 Jm^{-2}	5.13 GPa
fargesii	Adult	4.41 MPa	1082 Jm^{-2}	5.7 GPa

Data from Foping suggests *F. qinlingensis* and *B. fargesii* are less tough and hard than the bamboo tested in this study, but have a higher Young's modulus. An explanation of why the modulus of these bamboos is higher could be that both the "young" and "adult" bamboo from Foping are actually older than the bamboo tested in this study. The "young" shoots tested in Foping are about 1 years old, but neither the young nor adult plants tested in this study exceed 8 months of age. Because of the speed at which bamboo can grow, it can reach its adult height of 3-30 meters in only a few months (Liese and Weiner, 1996). While these bamboos may have been of similar height, older bamboo may be stiffer than the younger plants due to changing chemical composition. Lignin is not deposited until after the first month in the bamboo growth cycle and cell wall thickening is known to continue at least until the end of the second year (Alvin and Murphy, 1988).

Material properties of giant bamboo (*Cathariostachys madagascariensis*) found on Madagascar have been studied previously by Yamashita *et al.* (2009) in an attempt to learn more about how bamboo lemurs process bamboo during feeding. The results of the lemur study are similar to the material properties found by this study. They report an outer culm (stem) toughness value of 8311 Jm^{-2}, a hardness value of 6.84 MPa for bamboo stems, and a Young's modulus of 9418 MPa. The bamboo in this study has a stem toughness of 6000-9000 Jm^{-2}, a hardness of 11-15 MPa, and a Young's modulus of 1-1.5 GPa (1000-1500 MPa). Although the overall hardness of bamboo from this study is higher, some individual specimens fell within range of the findings of Yamashita *et al.*, (2009). Differences in hardness and modulus could be caused by the age and species of the plant, the position on the plant the specimen was taken, and the technique used to estimate mechanical properties. The size of the bamboo culm (stems) studied by Yamashita *et al.* (2009) range from 15 to 60 mm in diameter while the bamboo from this study is much smaller, with a diameter of 3-5 mm for young plants and 10 -15 for adults. Both studies used identical techniques for obtaining toughness and hardness of stems, but the lemur study used a 3-point bending technique as opposed to the 4-point bend used here. The advantage of the 4-point bending test is that it is easier to perform and interpret results at the expense of more time spent preparing the specimen which can be difficult in the field (Lucas, 2004).

Lucas (2004) has published data on the mechanical properties of various materials such as leaves, seed coverings, animal fibers, and certain inorganic material. From these data, quartz is found to have a hardness of over 7000 MPa while having a toughness value of only 2 Jm^{-2}. In contrast, bamboo from this study has a hardness of around 11-15 MPa, but a stem toughness of 6000-9000 Jm^{-2}. This indicates bamboo is much tougher than it is hard which is consistent with the idea that mechanical defenses of organic materials are either hard or tough, but usually not both (Lucas, 2004). For comparisons with leaves of other plants, the leaves of *Castsanopsis fissa* (in the beech family, Fagaceae) have a toughness of 410 Jm^{-2} with a toughness of 2000-6000 Jm^{-2} across the veins and midrib. Bamboo leaves have a toughness of 1200-2500 Jm^{-2} and are also toughest across the midrib. For most bamboo leaves in this study, the midrib was found to have a toughness of about 500 Jm^{-2} greater than the surrounding tissue.

COMPARISON OF BAMBOO TO OTHER FOODS

A study on the toughness of common foods eaten by mountain gorillas (*Gorilla gorilla beringei*) has reported several toughness values for foods that make up a significant proportion of gorilla diets (Elgart-Berry, 2004). The toughest foods listed in the study include the bark of *Ficus natalensis*, and *Eucalyptus* trees as well as the bark of the shrub *Piper capenesis*. These materials were reported to have toughness values of 4000 to 6000 Jm^{-2}. These values fall within the range of the bamboo stems tested by this study. The tree and shrub bark with the highest toughness values are not the most common food items selected by mountain gorillas and make up only 1-3% of their diet (Elgart-Berry, 2004). More common foods include the stems of the herb *Carduus afromontanis* (toughness of 1910 Jm^{-2}) and various fruits which range from 20 to 1100 Jm^{-2}. Mountain gorillas also consume the leaves of various trees and shrubs which vary greatly in toughness from about 20 to 1200 Jm^{-2}. Some of the toughest leaves and fruits are comparable to the toughness values found for bamboo leaves. Therefore, giant pandas likely place a much greater amount of stress on their jaws and teeth than mountain gorillas as giant pandas feed almost constantly on tough bamboo.

Table 2. Mechanical properties of bamboo compared to other organic and inorganic materials. The table below compares bamboo mechanical properties found in this study to properties found in previous research. Bamboo properties list below refer to the culm of the plant unless otherwise specified as (leaf). Constantino, 2006 refers to unpublished results.

Material	Toughness	Hardness	Young's Modulus	Source
Pseudosasa japonica	5625 ± 1570 Jm^{-2}	11.0 ± 2.4 MPa	1.3 ± 0.67 GPa	King, 2014
Phyllostachys bissetii	6696 ± 1791 Jm^{-2}	13.9 ± 4.5 MPa	1.5 ± 0.59 GPa	King, 2014
Phyllostachys dulcis	5758 ± 2280 Jm^{-2}	9.8 ± 2.9 MPa	1.1 ± 0.51 GPa	King, 2014
Phyllostachys nigra	9662 ± 2450 Jm^{-2}	8.6 ± 3.1 MPa	1.2 ± 0.20 GPa	King, 2014

P. japonica (leaf)	2015 ± 795 Jm^{-2}			King, 2014
P. bissetii (leaf)	1383 ± 526 Jm^{-2}			King, 2014
P. dulcis (leaf)	1186 ± 234 Jm^{-2}			King, 2014
P. nigra (leaf)	2186 ± 1098 Jm^{-2}			King, 2014
Fargesia qinlingensis (young)	508 Jm^{-2} (leaf)	1.1 MPa	3.4 GPa	Constantino, 2006
F. qinlingensis (adult)	366 Jm^{-2} (leaf)	2.82 MPa	7.8 GPa	Constantino, 2006
Bashaina fargesii (young)	864 Jm^{-2} (leaf)	1.76 MPa	5.13 GPa	Constantino, 2006
B. fargesii (adult)	1082 Jm^{-2} (leaf)	4.41 MPa	5.7 GPa	Constantino, 2006
Cathariostachys madagascariensis	8311 Jm^{-2}	6.84 MPa	9.4 GPa	Yamashita *et al.*, 2009
Non Bamboo Materials				
Material	Toughness	Hardness	Young's Modulus	Source
Quartz	2 Jm^{-2}	7000 MPa		Lucas, 2004
Castsanopsis fissa (leaf)	410 Jm^{-2}			Lucas, 2004
Eucalyptus (bark)	5430 Jm^{-2}			Elgart-Berry, 2004
Rhizome	5448 Jm^{-2}		1.1 GPa	Dominy *et al.*, 2008
Tuber	1304 Jm^{-2}		0.5 GPa	Dominy *et al.*, 2008

OTHER BAMBOO FEEDERS

Mountain gorillas, in addition to eating tough tree barks, have also been documented eating bamboo (Elgart-Berry, 2004). Despite being capable of masticating tough materials, the bamboo eaten by mountain gorillas are very young shoots which are low in toughness (the bamboo species *Arundinaria alpine* has a toughness of about 190 Jm^{-2}), Elgart-Berry (2004) reported that the bamboo consumed by mountain gorillas was not woody in consistency, unlike the bamboo tested in this study. Lemurs in the genus *Hapalemur* are bamboo specialists despite previous attempts to link bamboo consumption with large body size (Schaller, 1963). In bamboo lemurs, the tooth size and shape seems to be suited for puncturing and crushing bamboo which allows them to process their selected food item despite their relatively small size (Seligsohn and Szalay, 1978). In the same study, Seligsohn and Szalay (1978) describe the width and rigidity of the stem as the limiting factors of bamboo consumption. Yamashita *et al.* (2009) have described the method in which the bamboo lemur (*Hapalemur simus*) circumvents the problem of bamboo not fitting between the upper and lower jaws. *H. simus* grips the bamboo with its hands and uses its upper canines and lower premolar to puncture the bamboo culm at the hollow internode space. After a hole has been made, *H. simus* strips away the outer culm to get at the inner culm pith which Yamashita *et al.*, (2009) reports is less tough than the outer culm (5800 Jm^{-2} rather than 8000 Jm^{-2}) and is made less tough through peeling instead of cutting with the teeth (as low as 400 Jm-2). Giant pandas do not need to form a hole before exposing the inner pith and can crack bamboo at the widest point between its upper and lower molars and peel back the culm with their teeth (Dierenfeld *et al.*, 1982).

The red or lesser panda (*Ailurus fulgens*) is not only a bamboo specialist, but also shares anatomical characters with the giant panda which are useful for bamboo mastication (Figueirido *et al.*, 2012). These characters include a shortened snout length, a shortened braincase, broad zygomatic arches, and enlarged molars with comparatively reduced canines (Figueirido *et al.*, 2010; Figueirido *et al.*, 2012). The researchers attribute the convergent morphology of red and giant pandas to the selective pressures of bamboo mastication. This is evidenced by the reasoning that the shared traits are unlikely to have been derived from a common ancestor because fossil evidence indicates that giant pandas and red pandas are not closely related (Salesa *et al.*, 2006). Salesa *et al.* (2006) states the false thumb, which is now used for bamboo manipulation in both pandas, was derived independently and was once used by ancestors of the red panda to aid in arboreal locomotion. It should be noted, however, that red pandas eat only the leaves and very young shoots of bamboo that are not yet woody in consistency (Wei *et al.*, 1999). The researchers contrast this with the giant panda which utilizes almost every part of the plant. This difference in bamboo feeding behavior could be caused by the difference in size of the two pandas. Although the dentition of the red panda may allow for higher bite forces (Figueirido *et al.*, 2012), its relatively small size may still make the mastication of the tougher bamboo stems difficult.

The similar masticatory morphology of giant and red pandas lends credence to the idea that dietary preference can drive evolutionary adaptions for consuming said diet and that shared morphology may be useful when inferring what foods may be eaten. If the properties of the bamboo found in this study are indicative of the bamboo which possibly drove the specialized anatomy of giant pandas, then perhaps foods with similar properties were responsible for the evolution of robust crania in *Paranthropus*.

RELEVENCE TO GIANT PANDA FEEDING

Of the four bamboo species tested, the giant panda preferred *P. japonica* over both *P. nigra* and *P. bissetii* as a food source in a study on bamboo preference of giant pandas (Tarou *et al.*, 2005). This bamboo was not found to be the hardest or toughest of the species that were examined. *P. japonica* differs morphologically from both *P. nigra* and *P. bissetii* in that it has larger leaves than either species (Tarou *et al.*, 2005; Unpublished Observations). Leaves on *P. japonica* also branch off from a single rachis rather than splitting off from several smaller branches (See figure below). Because the leaves grow on a single rachis, this may make it easier for giant pandas to eat. Dierenfeld et al., (1982) describes the technique giant pandas use to eat leaves. They grasp the stem and place it in their teeth, then pull the stem away from them while twisting their neck in the opposite direction. Having all the leaves on one rachis may make it quicker and easier for giant pandas to eat all of the leaves on one shoot. The morphological characters of *P. japonica* could be what make it more attractive as a food source than the other tested bamboos.

P. japonica   P. nigra   P. bissetii

Figure 1. The Relative Size and Structure of P. japonica *as Compared to* P. nigra *and* P. bissetii. *The leaves of* P. japonica *are larger than the other species and originate from a single rachis, rather than several smaller branches.*

Of the three species of bamboo used in a study of bamboo preference in giant pandas (Tarou et. al., 2005), black bamboo (*P. nigra*), while still acceptable for consumption, was the least preferred species. While this preference may be a matter of smaller vs. larger leaf size, the findings of this study show *P. nigra* to be the toughest of these species in both leaves and stems. The extra toughness of *P. nigra* may be enough to dissuade giant pandas from feeding on it when a less tough alternative is available. Dierenfeld *et al.* (1982) has reported that leaves are the most digestible part of bamboo for giant pandas, but bamboo part preference varies throughout the year and leaves are only consumed from midsummer to winter with shoots and culm being preferred in the spring (Wei *et al.*, 1999; Hanson *et al.*, 2010). Because leaves are not consumed year-round, it seems unlikely that leaf size should play a significant role in the food selection of giant pandas. If leaf size is not the reason for preference of *P. japonica*, the toughness of the bamboo stems may be responsible for making *P. nigra* a less attractive food source for giant pandas.

POSSIBLE FOOD SOURCES OF *PARANTHROPUS*

Because of the robust cranial features of *Paranthropus*, it was long assumed that it relied on a diet of hard nuts or seeds, using its powerful jaws and teeth to crack open hard food objects (Tobias, 1967). However, the efforts of Cerling *et al.* (2011) and Ungar *et al.* (2008) have combined to reveal the diet of *Paranthropus* to contain high amounts of C_4 plant material and microwear patterns which show no evidence of the consumption of hard food objects. While this evidence is in stark contrast to the idea that *Paranthropus* was using its teeth to crack nuts and seeds, these results do not agree with *Paranthropus's* functional morphology (Constantino and Wood, 2007). The large, cusped postcanine teeth as well as the lack of high shearing crests used to process fibrous leaves and plant material indicate *Paranthropus* was probably not eating many grasses (Kay, 1975). While hard object feeding in *Paranthropus* now seems unlikely, Laden and Wrangham (2005) have proposed that underground storage organs (USOs) may not only have been a fallback food for *Paranthropus*, but perhaps even a preferred food source. They hypothesize that consuming raw USOs would have required an extensive amount of chewing. The high volume of chewing could possibly be a factor in the development of the derived morphology of *Paranthropus*.

A study by Dominy *et al.* (2008) has quantified the toughness and Young's modulus of a variety of USOs. This study reports bulbs and corms to be the least tough (around 300 Jm^{-2}) and have Young's modulus of 2 to 5 MPa. While both values are low compared to what is reported for bamboo, rhizomes and tubers were found to have toughness values comparable to bamboo (5400 and 1300 Jm^{-2}, respectively). If USOs such as tubers and rhizomes were an integral part of the diet of *Paranthropus*, the toughness of the USOs as reported by Dominy *et al.*, (2008) may have been a sufficient selective pressure for the adaptation of robust cranial features. The regular consumption of grass rhizomes could have also contributed to the high C_4 signal found in *Paranthropus*. However, Dominy *et al.* (2008) has deemed the consumption of rhizomes by *Paranthropus* to be unlikely, citing the tendency of human and extant apes that chew these rhizomes to ultimately eject them from the mouth. They instead offer the suggestion that tubers are a more likely food source as they are less tough than rhizomes and are similar in toughness to fruit tissue found in the diet of some apes. This suggestion does not address the C_4 conundrum as tubers do not typically utilize the C_4 pathway (Sage and Monson, 1999). Although rhizomes are significantly tougher than tubers, *Paranthropus* was likely able to generate higher bite forces than other hominins (Demes and Creel, 1988) and may have possibly been able to tolerate the higher toughness of rhizomes.

51

CONCLUSIONS

This study concludes that bamboo utilized by the giant panda for nutrition is tough, but not relatively hard. *Phyllostachys nigra* is both the toughest and least hard of the tested bamboos which is consistent with the idea that materials specializing in one form of mechanical defense are usually deficient in the other (i.e., hard materials are usually brittle and tough materials are easier to puncture). The specialized anatomical features of giant pandas make masticating tough bamboo possible, despite digestive anatomy which is ill suited to processing this unusual diet.

Much about the diet of *Paranthropus* still remains unknown, but the giant panda may be a useful model for uncovering those secrets. Perhaps *Paranthropus* and the giant panda are similar in the sense that they both consumed large quantities of nutrient poor foods in order to satisfy their metabolic needs. More research into how tough and hard food consumption help select for cranial morphology would be beneficial to understanding the diet of *Paranthropus*. It is currently unknown how large masticatory muscles can be differentiated between processing hard or tough materials. In the case of the giant panda, it seems its powerful jaws are suited to repetitive chewing and not so much the cracking of hard objects. Perhaps a similar case can be made for *Paranthropus*.

REFERENCES

Alvin, K., & Murphy, R. J. (1988). VARIATION IN FIBRE AND PARENCHYMA WALL TlfiCKNESS IN CULMS OF THE BAMBOO SINOBAMBUSA TOOTSIK.

Ambrose, S. H. (2006). A tool for all seasons. *SCIENCE-NEW YORK THEN WASHINGTON-, 314*(5801), 930.

Cerling, T. E., Mbua, E., Kirera, F. M., Manthi, F. K., Grine, F. E., Leakey, M. G., ... & Uno, K. T. (2011). Diet of Paranthropus boisei in the early Pleistocene of East Africa. *Proceedings of the National Academy of Sciences, 108*(23), 9337-9341.

Christiansen, P. (2007), Evolutionary implications of bite mechanics and feeding ecology in bears. Journal of Zoology, 272: 423–443. doi: 10.1111/j.1469-7998.2006.00286.x

Christiansen, P., & Wroe, S. (2007). Bite forces and evolutionary adaptations to feeding ecology in carnivores. *Ecology, 88*(2), 347-358.

Choong, M. F., Lucas, P. W., Ong, J. S. Y., Pereira, B., Tan, H. T. W., & Turner, I. M. (1992). Leaf fracture toughness and sclerophylly: their correlations and ecological implications. *New Phytologist, 121*(4), 597-610.

Constantino PJ (2007) Primate Masticatory Adaptations to Fracture-Resistant Foods. Ph.D. (The George Washington University, Washington, D.C.).

Constantino, Paul J., Peter W. Lucas, James J. W. Lee, and Brian R. Lawn. 2009. The influence of fallback foods on great ape tooth enamel. *American Journal of Physical Anthropology* 140:653–660.

Daegling DJ, McGraw WS, Ungar PS, Pampush JD, Vick AE, et al. (2011) Hard-Object Feeding in Sooty Mangabeys (Cercocebus atys) and Interpretation of Early Hominin Feeding Ecology. PLoS ONE 6(8): e23095. doi:10.1371/journal.pone.0023095

Darvell, B. W., Lee, P. K. D., Yuen, T. D. B., & Lucas, P. W. (1996.). A portable fracture toughness tester for biological materials . Measurement Science and Technology, 7(6), doi: doi:10.1088/0957-0233/7/6/016

Davis, DD. (1964), The giant panda: a morphological study of evolutionary mechanism. Fieldiana Zoology Memoirs 3, 1-339.

Demes, B., & Creel, N. (1988). Bite force, diet, and cranial morphology of fossil hominids. *Journal of Human Evolution*, *17*(7), 657-670.

Dierenfeld ES, Hintz HF, Robertson JB, Van Soest PJ, Oftedal OT. (1982). Utilization of bamboo by the giant panda. J Nutr. Apr;112(4):636-41.

Domínguez-Rodrigo, M., Pickering, T. R., Baquedano, E., Mabulla, A., Mark, D. F., Musiba, C., ... & Arriaza, M. C. (2013). First partial skeleton of a 1.34-million-year-old Paranthropus boisei from Bed II, Olduvai Gorge, Tanzania. *PloS one*, *8*(12), e80347.

Dominy, N., Vogel, E. R., Yeakel, J. D., Constantino, P., Lucas, P. W., 2008 Mechanical Properties of Plant Underground Storage Organs and Implications for Dietary Models of Early Hominins Evol. Biol. 35, 159.

Du Brul, E. L. (1977), Early hominid feeding mechanisms. Am. J. Phys. Anthropol., 47: 305–320. doi: 10.1002/ajpa.1330470211

Elgart-Berry, A. (2004). Fracture toughness of mountain gorilla (Gorilla gorilla beringei) food plants. *American Journal of Primatology*, *62*(4), 275-285.

Figueirido, B., SERRANO-ALARCÓN, F. J., Slater, G. J., & Palmqvist, P. (2010). Shape at the cross-roads: homoplasy and history in the evolution of the carnivoran skull towards herbivory. *Journal of evolutionary biology*, *23*(12), 2579-2594.

Figueirido, B., Serrano-Alarcón, F. J., & Palmqvist, P. (2012). Geometric morphometrics shows differences and similarities in skull shape between the red and giant pandas. *Journal of Zoology*, *286*(4), 293-302.

Grine FE, Martin LB. (1988). Enamel thickness and development in Australopithecus and Paranthropus. In: Grine FE, editor. Evolutionary history of the "robust" Australopithecines. New York: Aldine de Gruyter. p 3–42.

Grine, F. E., Ungar, P. S. & Teaford, M. F. 2006a Was the Early Piocene hominin 'Australopithecus' anamensis a hard object feeder? S. Afr. J. Sci. 102, 301–310.

Grine, F. E., Ungar, P. S., Teaford, M. F. & El-Zaatari, S. 2006b Molar microwear in Praeanthropus afarensis: evidence for dietary stasis through time and under diverse paleoecological conditions. J. Hum. Evol. 51, 297–319. (doi:10.1016/j.jhevol.2006.04.004)

Hirayama, K., Kawamura, S., Mitsuoka, T. and Tashiro, K. (1989), The faecal flora of the giant panda (Ailuropoda melanoleuca). Journal of Applied Microbiology, 67: 411–415. doi: 10.1111/j.1365-2672.1989.tb02511.x

Jain, S., Kumar, R., Jindal, U. C. (1992). Mechanical behavior of bamboo and bamboo compostite. *Journal of Materials Science*, 27, 4598-4604.

Kay, R. F. (1975). The functional adaptations of primate molar teeth. *American Journal of Physical Anthropology*, *43*(2), 195-215.

Kay, R. F. (1985). Dental evidence for the diet of Australopithecus. Annual Review of Anthropology, 14, 315–341. doi:10.1146/annurev.an.14.100185.001531

Kay, R. F., & Covert, H. H. (1984). Anatomy and behaviour of extinct primates. In *Food acquisition and processing in primates* (pp. 467-508). Springer US.

Kobayashi, T., Ikeda, H. and Hon, Y. (1999), Growth Analysis and Reproductive Allocation of Japanese Forbs and Grasses in Relation to Organ Toughness under Trampling. Plant Biology, 1: 445–452. doi: 10.1111/j.1438-8677.1999.tb00727.x

Laden, G., & Wrangham, R. W. (2005). The rise of the hominid as an adaptive shift in fallback foods: Plant underground storage organs (USOs) and austalopith origins. *Journal of Human Evolution*, 49, 482–498.

Li, Z., & Denich, M. (2004). Is Shennongjia a suitable site for reintroducing giant panda: an appraisal on food supply. *Environmentalist, 24*(3), 165-170.

Liem KF (1990). Aquatic versus terrestrial feeding modes: Possible impacts on the trophic ecology of vertebrates. *American Zoologist* 30, 209–21.

Liese, W. & G. Weiner. 1996. Ageing of bamboo culms. A review. *Wood Sci. Technol.* 30: 77–89.

Low,I.M, Che, and Latella, B.A. (2006). Mapping the structure, composition and mechanical properties of bamboo. Journal of Materials Research, 21: 1969-1976. doi:10.1557/jmr.2006.0238.

Lucas, P, Turner, I, Dominy, N, and Yamashita, N. (2000), Mechanical Defenses to Herbivory. *Annals of Botany.* 86(5): 913-920 doi:10.1006/anbo.2000.1261

Lucas, P. W. (2004). Dental functional morphology: How teeth work. Cambridge: Cambridge University Press.

Lucas, P., Constantino, P., Wood, B., & Lawn, B. (2008). Dental enamel as a dietary indicator in mammals. *BioEssays, 30*(4), 374-385.

Lucas, Peter W., Zhongquan Sui, Kay Y. Ang, Hugh Tiang, Wah Tan, Sheau H. King, Brooke Sadler, and Neeraja Peri. (2009). Meals versus snacks and the human dentition and diet during the Paleolithic. In *The evolution of hominin diets: integrating approaches to the study of Palaeolithic subsistence.* Jean-Jacques Hublin and Michael P. Richards, eds. Pp. 31–41. Dordrecht: Springer.

McClure, F. A. (1993). The bamboos. (pp. 28-42). Boston: Smithsonian Institution.

Olejniczak, A. J., Smith, T. M., Skinner, M. M., Grine, F. E., Feeney, R. N., Thackeray, J. F., & Hublin, J. J. (2008). Three-dimensional molar enamel distribution and thickness in Australopithecus and Paranthropus. *Biology Letters, 4*(4), 406-410.

Robinson, B. W., & Wilson, D. S. (1998). Optimal foraging, specialization, and a solution to Liem's paradox. *The American Naturalist, 151*(3), 223-235.

Rybiski Tarou, L., Williams, J., Powell, D. M., Tabet, R., & Allen, M. (2005). Behavioral preferences for bamboo in a pair of captive giant pandas (Ailuropoda melanoleuca). *Zoo Biology, 24*(2), 177-183.

Sage, R. F., & Monson, R. K. (1999). C4 plant biology. New York: Academic Press.

Salesa, M. J., Antón, M., Peigné, S., & Morales, J. (2006). Evidence of a false thumb in a fossil carnivore clarifies the evolution of pandas. *Proceedings of the National Academy of Sciences of the United States of America, 103*(2), 379-382.

Schaller, G. E. (1963). The mountain gorilla: Ecology and behavior.

Seligsohn, D., & Szalay, F. S. (1978). Relationship between natural selection and dental morphology: tooth function and diet in Lepilemur and Hapalemur. *Studies in the development, function and evolution of teeth (PM Butler and KA Joysey, eds.). Academic Press, London, United Kingdom*, 289-307.

Smith, J. W., & Walmsley, R. (1959). Factors affecting the elasticity of bone. *Journal of anatomy, 93*(Pt 4), 503.

Spencer, M. A. (1998). Force production in the primate masticatory system: electromyographic tests of biomechanical hypotheses. *Journal of human evolution, 34*(1), 25-54.

Sponheimer, M., Lee-Thorp, J., de Ruiter, D., Codron, D., Codron, J., Baugh, A. T., et al. (2005b). Hominins, sedges, and termites: New carbon isotope data from the Sterkfontein valley and Kruger National Park. Journal of Human Evolution, 48, 301–312. doi:10.1016/j.jhevol.2004.11.008.

Tobias PV. (1967). The cranium and maxillary dentition of Australopithecus (Zinjanthropus) boisei. OlduvaiGorge, Vol. 2. Cambridge: Cambridge University Press.

Turner, I. M., Choong, M. F., Tan, H. T. W., & Lucas, P. W. (1993). How tough are sclerophylls?. *Annals of Botany*, *71*(4), 343-345.

Ungar, P. S. 2004 Dental topography and diets of Australopithecus afarensis and early Homo. J. Hum. Evol. 46, 605–622. (doi:10.1016/j.jhevol.2004.03.004)

Ungar, P. S., Grine, F. E. & Teaford, M. F. 2008 Dental microwear indicates that Paranthropus boisei was not a hard-object feeder. PLoS ONE 3, 1–6.

Ungar, P. and Sponheimer, M. (2011), The diets of early hominins. Science, 334: 190-193. doi: 10.1126/science.1207701

Ungar, P. S. (2012). Dental Evidence for the Reconstruction of Diet in African Early Homo. *Current Anthropology*, *53*(S6), S318-S329.

van der Merwe, Nikolaas J.; Masao, Fidelis T. and Bamford, Marion K. (2008). Isotopic evidence for contrasting diets of early hominins Homo habilis and Australopithecus boisei of Tanzania. S. Afr. j. sci. 104: 153-155.

Walker, A. (1981). Dietary hypotheses and human evolution. *Philosophical Transactions of the Royal Society of London. B, Biological Sciences*, *292*(1057), 57-64.

Wei, F., Feng, Z., & Wang, M. (1999). Feeding strategy and resource partitioning between giant and red pandas. Mammalia, 63(4), 417-430. doi: 10.1515/mamm.1999.63.4.417

Williams, C. L., Willard, S., Kouba, A., Sparks, D., Holmes, W., Falcone, J., Williams, C. H. and Brown, A. (2012), Dietary shifts affect the gastrointestinal microflora of the giant panda (*Ailuropoda melanoleuca*). *Journal of Animal Physiology and Animal Nutrition*. doi: 10.1111/j.1439-0396.2012.01299.x

Wood, Bernard, and David Strait. 2004. Patterns of resource use in early *Homo* and *Paranthropus*. *Journal of Human Evolution* 46:119–162.

Wood, B., & Constantino, P. (2007). *Paranthropus boisei*: Fifty years of evidence and analysis. *Yearbook of Annual Anthropology, 50*, 106-132.

Yamashita, N., Vinyard, C. J., & Tan, C. L. (2009). Food mechanical properties in three sympatric species of Hapalemur in Ranomafana National Park, Madagascar. *American journal of physical anthropology, 139*(3), 368-381.

Yu, G., Jiang, Z., Zhao, Z., Wang, B. and Wang, Y. (2003), Feeding habitat of giant pandas (*Ailuropoda melanoleuca*): why do they prefer bamboo patch edges?. Journal of Zoology, 261: 307–312. doi: 10.1017/S0952836903004242

APPENDIX A: LETTER FROM INSTITUTIONAL RESEARCH BOARD

MARSHALL
UNIVERSITY®
w w w . m a r s h a l l . e d u

Office of Research Integrity

May 6, 2013

Ryan King
513 Camp Hill Rd.
Elizabeth, WV 26143

Dear Mr. King:

This letter is in response to the submitted thesis abstract titled "Mechanical Properties of Bamboo and Ecomorphology of *Ailuropoda melanoleuca*." After assessing the abstract it has been deemed not to be human subject research and therefore exempt from oversight of the Marshall University Institutional Review Board (IRB). The Code of Federal Regulations (45CFR46) has set forth the criteria utilized in making this determination. Since the information in this study does not involve human subjects as defined in the above referenced instruction it is not considered human subject research. If there are any changes to the abstract you provided then you would need to resubmit that information to the Office of Research Integrity for review and a determination.

I appreciate your willingness to submit the abstract for determination. Please feel free to contact the Office of Research Integrity if you have any questions regarding future protocols that may require IRB review.

Sincerely,

Bruce F. Day, ThD, CIP
Director
Office of Research Integrity

Printed by Books on Demand GmbH, Norderstedt / Germany